# Chronicles of a Gorkha Soldier

# Chronicles of a Gorkha Soldier

By

Brig R V Singh

Vij Books India Pvt Ltd
New Delhi (India)

*Published by*

**Vij Books India Pvt Ltd**
(Publishers, Distributors & Importers)
2/19, Ansari Road
Delhi – 110 002
Phones: 91-11-43596460, 91-11-47340674
Mob: 98110 94883
e-mail: contact@vijpublishing.com
web : www.vijbooks.in

First Published in India in 2022



ISBN: 978-93-93499-36-3

# Contents

## PART – II
## COMMAND AND STAFF

# INTRODUCTION

Actually, writing a book on my experiences during service in Army has never been the intention. It all has been kind courtesy insistence by numerous well wisher friends, my wife and son who emphasized I pen down experiences of my 35 years long service in Army and a decade thereafter under civil dispensation. I had been fortunate to participate in the 1971 Indo-Pak War as a 2/Lieutenant with just one year of service and led a Rifle Platoon in an assault by my Battalion on an enemy stronghold, named SEHJRA in KHEM-KARAN Sector of Punjab. It has been a great blessing to be with my Battalion in almost all operational areas during my formative years of service; 2/Lieutenant in J&K and during 1971 Indo- Pak War – a once in life time opportunity which every soldier does not get in his entire service, Captain in Arunachal Pradesh and during heightened insurgency in Mizoram during 1973-76, had two tenures in high altitude area in Arunachal Pradesh as Major and Colonel commanding a Rifle Company and an Infantry Battalion in the mid 1980s and early 1990s respectively. Also, I happened to do a string of Staff appointments from the rank of a Captain to Brigadier in almost all formation HQ i.e. Brigade, Division, Corps, Command and Sub Area, and an Instructional appointment in Infantry School. Attendance at Staff Course at DSSC, Wellington (TN) and Naval Higher Command Course at College of Naval Warfare, Mumbai were good exposures during service career. All these avenues gave me an excellent opportunity to interact with a wider spectrum of people and be witness to certain incidents unfolding before me which otherwise would not have been possible.

This book is an attempt to record some of the incidents which I personally underwent or observed them from close vicinity in the course my service in Army and in civil street, jointly lasting for almost half a century. An honest effort has been made to present these incidents before the readers as they unfolded before me at varied place and time, in a truthful, impartial and unbiased manner, as a direct participant or a close witness to each of them.

These Chronicles, as the heading suggests, list out the experiences of an Officer who is commissioned as a 2/Lieutenant in a Gorkha Regiment, moves with his Battalion for next 7 - 8 years continuously where ever it goes. Battalion is his HOME for next 20 - 22 years. He lives with his Soldiers, eats, plays, interacts and trains with them. His soldiers are everything to him. With an aim to know my Soldiers better, their religion, culture, social customs and traditions in detail and at the earliest, I visited their homes in Nepal 5 times during my annual leaves and furloughs lasting 15 days to 4 months at a stretch and at my own risk and cost. It is simply great to undertake a few months long trek in higher reaches of Himalayas of Central Nepal just to meet a Soldier of your Battalion who has come on Annual Leave and his village could well be 10 days walking distance from the nearest Road Head. The happiness which one gets on seeing the bewildered face of the Soldier when he notices a 2/Lieutenant from his Battalion standing at the door of his house in a village, deep in the interior of the remote part of barren high mountains cannot be explained; same can only be experienced. Such an effort is always well worth it and was certainly so in my case. Those are still the most cherished memories of my entire 35 years of service in Army. It may sound as an overdose of REGIMENTATION to few but it was not so. In fact it established me amongst my Soldiers and won their heart and trust, a true reward for genuine REGIMENTATION.

I happened to come across an article in USI Journal of 1974 on REGIMENTATION written by Lt Col (later Maj Gen) S C Sardeshpande, KUMAON Regiment, while undergoing QUARTER MASTERS Course at AOC School, Jabalpur. He had covered this subject in a very methodical and practical manner and laid out a proper route for young officers to move ahead on the path of correct regimentation in an Infantry Battalion. I tried to imbibe the advices rendered by the writer and made a fair and determined effort to move ahead in my professional Life.

I have been fortunate to serve under professionally very competent, mature, caring and understanding Senior Commanders at all levels. In fact, many of them have been instrumental in extricating me from few not so happy situations. I am forever grateful to each one of them. My colleagues and subordinates have always supported me and achieved the impossible whenever the need arose. I am proud a Commanding Officer whose one Company Commander rose to become Chief of Army Staff, two became Major Generals, two Brigadiers and four Colonels ( all commanded Infantry Battalions ). I am always grateful to them.

This book has been divided in 4 parts as under:-

### PART – 1: REGIMENTATION - FORMATIVE YEAR

This part forms major part of the book and contains 21 articles out of total 44 articles and covers various incidents which took place during my formative years of service.

### PART – 2: COMMAND AND STAFF

This part also forms large part of the book and contains 18 articles covering the period spent on numerous command staff appointments.

### PART – 3: SOME PARANORMAL EXPERIENCES

This part contains three articles on some Paranormal Experiences which were experienced while working as Director Ex-servicemen Welfare with Madhya Pradesh Government and another at Hotel Savoy, Mussoorie.

### PART – 4: ADVENTURE AND SPIRITUALITY

This part contains three articles on adventure and spirituality. First article covers a Trek to Satopanth Tal, known as SWARGAROHINI or CLIMB TO HEAVEN Trek, beyond Shree Badrinath Dham. This

Trek was undertaken by mighty PANDAVAS of MAHABHARAT to wash away the sins of having killed 100 KAURVAS, their own brethren. Second article pertains to my visit to PIOUS KAILASH AND MANSAROVAR and third article gives details of witnessing a divine spectacle beyond Shree Badrinath Dham in the midst of high Himalayas.

For ease of understanding, this book has been written in anecdotal form and as a first person account of incidents and experiences which one has undergone while in Army service and thereafter. It should make a good leisurely reading for all.

# PART – I

# REGIMENTATION – FORMATIVE YEARS

# GYARAH SAHEB

## SUB MAJ / HONY CAPT NAR BAHADUR GURUNG

## ADC TO THE PRESIDENT OF INDIA

The first time I came across this name, 'GYARAH SAHEB' was on the very next day of joining the Battalion in September 1970. The battalion was located at Sarol, a small place between Pir Badesar and Rajauri in J&K. I met GYARAH SAHEB after an Inter-Company Football Match in which he had participated. He was a tall, well built, smart and respectful person and a Havildar those days. He always appeared a little different from others and had an aura of a good upbringing around him. This aspect was better understood by me two years later when I visited his home in Nepal.

He belonged to a village named BACHHA in PAIYUM district of Western Nepal. I went on my first two months long Trek to Regimental Recruiting Areas in September - October 1972 and visited the home of GYARAH SAHEB. By then, he had been promoted to the rank of Nb Sub. Not to be misunderstood by the uninitiated of the strange way of calling each other among JCO / OR in Gorkha Battalions, GYARAH denotes the last two digits of Army Number assigned to Nar Bahadur Gurung Saheb after attestation on completion of his basic recruit training at the Training Centre. It is a custom in all Gorkha Infantry Battalions that all soldiers are better known amongst themselves by the last two digits of their seven digit Army number. In case Rifleman Rup Bahadur Gurung has been assigned an Army Number as 5443761, he will be better known in the battalion as '61', throughout his Service and even after retirement too. It is not uncommon even the wife of a

Gorkha Soldier calling her husband by last two digits of his Army number and not by his actual name. In October-November 1974, I went on my 2nd trip, three months long this time, and visited a famous Gurung village named SIKLES, North of POKHARA in Nepal. A Havildar named Akal Bahadur Gurung had passed JCO Cadre and was waiting for promotion to the rank of Nb Sub. He was in the battalion when I visited his home in Nepal and met his wife. Very respectfully, she asked me, "HAMRO CHAUNTIS KAILE JEMADAR BANCHHA – WHEN WILL OUR THIRTY FOUR BECOME A JEMADAR?". Yes, her 34, Havildar Akal Bahadur Gurung , not only became a Jemadar but retired as a proud Hony Capt later. It is expected from all officers of a Gorkha Battalion to remember the maximum of the Soldiers of the Battalion by the last two digits of their Army Numbers. There would be many persons of the same common name in the Company, leave aside in the whole battalion. This helps in minimising the confusion as also establishes better rapport among officers and soldiers under their command. How does a Young Officer memorise these numbers? I had maintained a proper Note Book in which every numeral from 01 to 99 had been allocated a page each which contained the Name and Company of individuals whose personal numbers ended with these numerals. Then there were pages earmarked separately for 100, 200 and so on and were called as 'poorey' (complete). When I moved from one Rifle Company to other, I updated my Note Book with fresh details. After 2 -3 years, I could memorise these numbers for the majority of soldiers of the battalion and always took pride in doing so.

During my second visit to Nepal in 1974, I again happened to visit the village of GYARAH Saheb, stayed in his home for three days, made it as a centre point to visit other neighbouring village to interact with other retired and serving soldiers. This time he was also on annual leave during the same period and assisted me in my visits to neighbouring villages and interaction with numerous serving and retired soldiers of the Regiment. My interaction opened a different vista about him. It was during this visit to his village that I learnt of the immense amount of sacrifices made by three generations of this MOST ILLUSTRIOUS FAMILY for the battalion. My regards for him grew manifold.

His grandfather was Sub Maj of the Battalion prior to FIRST WORLD WAR, participated in numerous campaigns in North-West Frontier and awarded with SARDAR BAHADUR, MC, IDSM prior to his retirement. GYARAH SAHEB's father, JEMADAR DHANKAJI GURUNG died in a fierce hand to hand fight in the famous Battle of Cassino in Italy during SECOND WORLD WAR and was awarded MILITARY CROSS (Posthumously) in the battalion. His mother was heavily pregnant at the time of death of her husband. GYARAH SAHEB was born two months after the death of his father and was brought up by his dotting Mother and Grandmother. He had a huge house in his village and, in my then perception as a 2/Lt, it resembled more like a palatial Bungalow in an Army Cantonment, a FLAG STAFF HOUSE, shall I say!

As was a trend in his illustrious family, he got himself recruited at 58 GTC, Dehra Dun and was assigned to 8 GR (not 5 GR). Before his attestation in 58 GTC, his Grandmother came to 58 GTC all the way from interior of Nepal, met Commandant of the Centre and got her Grandson transferred to his ancestors' Regimental Battalion i.e. 1/5 GR. However, his Army Number, forever, remained of 8 GR and not 5 GR and started with 57 and not 54 as was the case in 5 GR. During my 3 days stay in their home in the village, his mother always waited for us to return from the neighbouring villages and had her meal after us. Eight poor boys from the neighbouring villages were regularly provided free accomodation, food, school fees and other misc expenses related to their education in the nearby High School at Bhoksing by his family. GYARAH SAHEB had been blessed with two daughters, elder one was married to an officer from Nepal Army, and younger one studying. Self and GYARAH SAHEB served in the battalion together for more than a decade, and I always held him in high esteem. He rose to become a JCO ADC to the President of India, third in a row from the battalion, and thus brought laurels to the battalion. He retired as Hony Capt in 1992 and settled down in POKHARA. I last met him in Pension Paying Office, Pokhara in June 1992. I had gone on visit to Nepal as CO 4/5 GR and he had joined PPO POKHARA on a special appointment.

What a proud lineage! What tremendous sacrifices made by the family! I am equally proud to have been associated with such GREAT SOLDIERS and always hold them in very high esteem. May GOD Almighty bless them all.

# I P SAHEB

## SUB MAJ /HONY CAPT INDRA PRASAD GURUNG

I joined 1/5 Gorkha Rifles on my commissioning as a 2/Lt when it was located at Sarol Camp near Rajouri in J&K in 1970. I was assigned to 'C' Company which had no Company Commander posted. It was here that I first met Subedar Indra Prasad Gurung or I P SAHEB, as he was generally known in the Battalion. He was Company Senior JCO and de facto Company Commander of 'C' Company, which I was posted in.

I P Saheb was a three generations old JCO in the Battalion. When I joined the Battalion, it was common to see many two and even three generations old Soldiers proudly strutting around. All these soldiers had an aura of generations long family respect and consequent responsibility, around their neck. They never demanded anything extra for the services and sacrifices rendered by their ancestors for the Battalion in the past and pre Independence era. It is unlike we officers who put in PARENTAL CLAIM as their right to command a specific Battalion; a JCO does not put up his claim for appointment as Sub Maj of the Battalion as PARENTAL CLAIM. 'I P Saheb' was a thoroughbred professional soldier, belonged to the 'old order' of soldiers and commanded tremendous respect in the Battalion. In fact, he was the last soldier, who did his basic recruit training in 56 Gorkha Rifles Centre at Abbotabad, now in Pakistan. He was extremely respectful to even a 2/Lt who barely had few days of Service. Always treated me like Officiating Company Commander, though he was actually carrying out all duties of the appointment. He ensured that CHM read out all orders/instructions received from Battalion HQ to me every day and organized a professional briefing of each of the Company appointment holders, like, CHM, CQMH, Company Clerk, KOTE NCO, Pay NCO, Cook House NCO to me in the first fortnight of my joining the Company.

'I P Saheb' kept a dutiful watch on me and fully well knew how I was utilising my spare time. After a month, he advised me not to spend much time in gossiping around in Officers' Mess and instead, I should read professional books from Officers Mess Library and GS Publications from Adjutant Office. He very respectfully advised me that as a good officer I should devote more time in reading professional books. Truly speaking, he acted like a 'father figure' to me and we both carried that relationship till he retired as a Sub Maj of the Battalion. He was quite senior in the hierarchy amongst JCOs and earmarked to take over as Sub Maj on retirement of then present incumbent Sub Maj Ganga Bahadur Rana, IOM. He was duly appointed as Sub Maj in June 1972.

'I P Saheb' had a long lineage of ancestors who served in the Battalion. His grandfather served in the Battalion and retired as a Subedar till the beginning of FIRST WORLD WAR. His father Sub Jit Bahadur Gurung served in the Battalion till SECOND WORLD WAR, was wounded in a battle and boarded out due to War Injury in the middle of War. His younger brother, Sub Hari Prasad Gurung, was serving as a MT JCO when I joined the Battalion and retired as a Hony Capt in late 1970s. His youngest brother served in the Battalion as an NCO, wounded in 1965 War in Khemkaran Sector and evacuated to Military Hospital. It may look strange; Reinforcement Camp, by a funny coincidence, posted him to 5/1 GR instead of 1/5 GR after recovery from the wound suffered in the battle. Later it was too late to rectify the mistake and he retired from 5/1 GR as a Havildar.

'I P Saheb' belonged to a village called DUNEDARA, near BHOKSING in PAIYUM District of Western Nepal. Paiyum area always had fairly large representation of JCOs and OR in our Battalion since long. I was rather fortunate to meet a large proportion, approx 20, serving and retired, persons of the Battalion during my short visit to this area. Numerous of these soldiers had participated even in FIRST WORLD WAR, leave aside SECOND WORLD WAR. It was here that I met a soldier from our Battalion who had retired as a Rifleman in 1922 and already earned his pension for half a century. It was truly a great feeling interacting with such BURHOS and listen to the stories of their grand association with the Battalion. I spent a full day and night at house of 'I P Saheb' in their village and in the company of his respected father. It was great talking to his father and listening to his experiences. 'I P Saheb' had a son and a daughter. His son, Driga Bahadur Gurung , a fourth-generation soldier, joined the Battalion in the mid-1970s and retired as Sub Major at 58 GTC. This is one of the MOST ILLUSTRIOUS FAMILY of the Battalion and their contribution more than a century old.

Out of his four years of tenure as Sub Maj, Battalion spent 1 year in Ferozepore and 3 years in Arunachal Pradesh / Mizoram. When we reached next Peace Station, Trivandrum, his tenure was almost over and he retired two months later. Before he finally left the Battalion, me and my wife, made a personal visit to his house in Trivandram, which was located quite close to our house and paid our regards to him and his wife. It was, indeed, a great occasion that I still cherish. I learnt so much from him in my younger days and still have great respect for him.

On retirement, 'I P Saheb' had comfortably settled down in his village. His son Driga Bahadur Gurung constructed a good house in Pokhara in mid-1990s and shifted his old father from village to Pokhara as his health was not keeping well. He breathed his last in mid-1990s at Pokhara. A truly great person, an affectionate human being, a thoroughbred professional soldier and a JCO of 'old order'. May God grant SADGATI to respected 'I P SAHEB'.

## ALE SAHEB

## SUB MAJ / HONY CAPT ARJUN SINGH ALE

## ADC TO THE PRESIDENT

My earliest memory of Sub Maj Arjun Singh Ale goes back to1970 when he was Company Senior JCO of 'D' Company while I had just joined the Battalion on commissioning from Officers' Training School, Madras and assigned to 'C' Company. Later, in mid-1972, I was shifted to 'D' Company where Ale Saheb was and that is how a life-long association between a 2/Lt and a Senior Subedar of the Battalion began.

Ale Saheb belonged to a place called Panditwadi in Dehra Dun. His father was a Subedar in 1/10 Gorkha Rifles. He was born in a Cantonment Town of Chaman, now in Pakistan, where his father's Battalion was stationed. He was Matriculate of 'old era' when even illiterate persons could also be recruited in various Gorkha Battalions. Strange as it may appear to many, numerous of the recruits, while undergoing basic training at 58 Gorkha Training Centre, could only converse in their local dialects, like, Gurung Kura, Magar Kura, Kham Kura etc and not even in Nepali language which was more of an intermediary language. Even during those days, Ale Saheb was a Matriculate amongst them and held with an awe by his colleagues. Professionally very competent and had high IQ. Also, he was very handsome, carried himself with dignity and always paid due respects to all. A mix of these qualities sometimes created jealousy amongst his peers and misunderstanding amongst a few Officers too. Actually, one has to know him properly to estimate his true worth. Take an example, I was going on 2 months leave in September-November 1972 to visit Regimental

Recruiting Areas in Nepal. Those days we had to either hire Bicycle from Unit Bania at a monthly rent of Rs 10/- or buy it, which cost Rs 100/- a hefty sum to pay those days. I had bought a Cycle for myself. Sub Ale Saheb requested me if I could leave the Cycle with him for these two months which I agreed. But many of my colleagues thought it was a smart act on the part of the JCO. Actually, it was never so; we had developed a good rapport which gave him that privilege to request.

He was very affectionate and respectful to me, a mere 2/Lt. He was ever ready to make me learn the intricacies of Company Administration and Man Management. What I had learnt from Sub Indra Prasad Gurung Saheb in 'C' Company two years ago was further refined in 'D' Company under the able guidance of Sub Arjun Singh Ale. He advised me to interview each and every Soldier of the Company with his IAFF-958 Service Record, Individual Pass Book, Training Records and personally verify each and every Occurrence / Entry and initiate necessary action to update. This was always carried out either in the afternoon or after dinner. It was a long drawn process involving 120 persons and each individual requiring at least 30 to 45 minutes for discussion and noting down the requirements. Follow up action needed many months-long continuous efforts at mine as well as Company Clerk's level. This laborious exercise established a perfect and everlasting rapport between me and my Company Soldiers.

In 1976, we lost Sub Bir Bahadur Gurung, our Sub Maj Designate, in a freak train accident while the Battalion was moving from Trivandrum to Pokharan for Desert Orientation Training. Sub Ale was next senior JCO but got overlooked and was posted out to 58 GTC Shillong. However, we continued to remain in touch. Destiny had better things lined for him. Soon, he was selected for the appointment of Sub Maj ADC to The President of India. He was definitely a befitting choice and held this august appointment for five years and functioned as JCO ADC to two Presidents i.e. Shree Neelam Sanjeeva Reddy and Giani Zail Singh. After retirement, he settled down at Delhi and not his home town Dehra Dun which surprised many including myself.

Sometime in 1979-80, he invited me and my wife to visit Rashtrapati Bhawan when Shree Neelam Sanjeeva Reddy was the President. He introduced us to a battery of ADCs from three Defence Services and other Civilian Services. We spent almost one and a half hour with him. Our seating arrangements were such that self and wife could well see the President sitting in his office and so could he. In fact, my wife could see the President even better than me, though she could not recognise him and thought just another person sitting in the adjoining room. After a while she nudged me and said that the person sitting in adjoining room is smiling at her. I had to quietly tell her that person giving her benevolent smile is none other than the President of India. Maybe that our visit that day as Guests of Ale Saheb had been conveyed to the President in advance.

We did not have the opportunity of meeting each other for the next almost four decades. I never forgot him, always thought of him, wished him well and always had sincere respect for him. I always wanted to meet him again and, with a great amount of running around, obtained his Delhi address; he was staying in Janakpuri with one of his sons named Anuj. I decided to meet and pay my regards to him in person. By now, he was in the early 90s of his life span. With a heart filled with emotions I reached his house and met him. I had so much to ask about happenings in his life during last four decades, so much to tell about my life since our last meeting in Rashtrapati Bhawan in 1979. Unfortunately, I found almost no visible reactions from him though he appeared fairly healthy for his advanced age. Alas! In actual fact, Ale Saheb had lost his memory two years ago and now recognised no one, not even his family members or his own house in the locality. Hence, he could not recognise me too. I felt sad and disheartened as I could not discuss anything with him. Yet I was much satisfied; at least, I could see him in person once again.

Sub Maj/Hony Capt Arjun Singh Ale had brought laurels to the Battalion. He was second JCO ADC to the President of India from the Battalion and in continuity as he had taken over from Sub Maj / Hony Capt Moti Prasad Thakali from the Battalion. Incidently, he handed over to another outstanding JCO from the Battalion, Sub Maj / Hony Capt Nar Bahadur Gurung. In fact one more JCO from the Battalion, Sub Maj / Hony Capt Tika Ram Thapa took over as JCO ADC to the President of India from Hony Capt Nar Bahadur Gurung, thus marking presence of a Sub Maj ADC from the Battalion in Rashtrapati Bhawan for 20 years in continuity.

It is so very true that the bond established with the troops in younger days is made forever and lasts forever. I do not know about others but can definitely say so for my own self. ALE SAHEB had desired to spend last days of his life at his ancestral house in Dehra Dun. His health worsened and became completely bed ridden. His son brought him to Dehra Dun where he breathed his last on 08 September 2021. May God grant SADGATI to the Soul of my dear CAPTAIN ARJUN SINGH ALE.

# UNKNOWN AND UNSUNG - JABAR BAHADUR GURUNG

A young 2/Lt is forever available to undertake any arduous physical activities, be that Long Range Patrols (LRP in short), Khud Race or participation in Gen Choudhary Trophy Competition of yesteryear. If one happens to join a Battalion in Field Area there are opportunities galore to lead umpteen numbers of Long Range Patrols in the higher reaches of the Himalayas. So was the case with me. I joined my Battalion 1/5 GR at a place called SAROL, near Rajauri in J&K in September 1970 and undertook two such Patrols in a span of 2 months, each of 10 days duration. It used to generally commence from the area of Mendhar / Krishna Ghati, climb on to Pir Panjal Ranges and get down somewhere near Thanamandi / Kalakot. I did one LRP in October and second in November 1970. It was great fun, physical fitness was not the problem and climbing mountains was desire of the heart. There could be nothing better.

It was in the second Patrol that I came to know of 5443024 Rifleman Jabar Bahadur Gurung. Patrol used to consist of a total strength of 12 to 14 persons. Each Rifle Company and Support Company used to detail 2 to 3 persons for the Patrol; Jabar Bahadur had been detailed from 'A' Company. I clearly remember his face, even till date, more than 52 years hence. He was short structured, like any Gorkha, fair looking, had number of small pox marks on his face and always smiling. His rucksack was always heaviest as Havildar Dharam Singh Gurung, Patrol Second-in-Command, always loaded him with lot of rations to carry and one odd cooking utensil dangling from some loose strap of his rucksack. He was very fit and always remained cheerful. Our Patrol had three persons from the Advance Party of 4/9 GR also as they were in the process of relieving our Battalion next month, December 1970.

Midway during the conduct of Patrol, we were required to link up with the Patrol of 6 SIKH coming from the area of Poonch to Pir Panjal Ranges. We climbed up on the given point on Pir Panjal and were soon contacted by the other Patrol. Within one minute of their arrival, and before we could even open our thermos flask to offer them tea, I saw a burly Khalsa standing in front of me, carrying a bucket which had a Kerosene Oil Stove burning and a Kettle placed on it. In no time, he lifted the Kettle, poured tea in a Mug (Enamelled) and offered that to me. I was surprised to see such quick action. This was the usual drill in any SIKH or PUNJAB Battalions to carry a Stove and Kettle in a bucket even while climbing mountains so as to ensure ready availability of tea to all ranks anytime.

After making formal contact with the other Patrol, we came down to little lower heights for the night halt. Often we used to make use of vacant huts of Bakkarwal for the night halt or else in some other suitable vacant place. Luckily we found a Bakkarwal Hut and occupied that for the night. We used to keep small fire burning the whole night either in the middle or in the close proximity of the Patrol to keep us warm as also drive away any wild animal, if any happen to come nearby. Night Sentry in any case was always on duty. During this night, a spark flew toward the edge of the Sleeping Bag in which Jabar Bahadur was sleeping. It was realized in the morning that the Sleeping Bag had been partially damaged by fire. Jabar Bahadur was quite worried of the damage and consequent recovery of the cost of Sleeping Bag from him. Those days salary used to be low and cost of items like Sleeping bag high. I had studied the scene of incident and was of the opinion that Jabar Bahadur was not to be blamed and assured him I would brief Battalion QM accordingly on our return from the Patrol and request him not make any deductions from the salary of Jabar Bahadur Gurung

We returned to the Battalion and I briefed Battalion Second-in-Command and QM of the damage caused to a Sleeping bag for which no one really was to be blamed. I spoke to Jabar Bahadur and again assured him of no recovery from him. Soon after, I was detailed on Young Officers' Course and rejoined the Battalion at its new location in Ferozepore in June 1971. I well remembered the incident and my assurance of 'No Recovery' given to Jabar Bahadur few months ago. Accordingly, I enquired from 'A' Company about this. It was sad to learn Jabar Bahadur was no more. He had met with an accident in March 71 with a Civil Truck and was crushed to death. It was further learnt he had been penalized for the damage caused to the Sleeping Bag and a deduction of Rs 300/- made from his IRLA Account. Those days Rs 300/- was a huge amount. Monthly salary of a Rifleman was around Rs 200/- pm. There were soldiers who used to draw just Rs 10/- in a month from their monthly Salary Account for the routine expenses during the month and save remainder for drawl while proceeding on Annual Leave. During 1970 – 71, a 2/Lt used to get a Rs 500/- monthly salary only. The worst part was the affected individual, Rifleman Jabar Bahdur Gurung was no more. If he was alive I could have compensated him and now even that option could not be exercised. I promised to myself to hand over this much amount to his dependents whenever I visit Nepal. Jabar Bahadur had not married till then and his old parents lived in the village in a district named SYANGJA in West Nepal.

An opportunity came my way soon. My request for proceeding on two months long visit to Regimental Recruiting areas in West Nepal during September – November 1972 was accepted by Army HQ. This gave me an opportunity to interact with the parents of late Jabar Bahadur and hand over the money to them, which I had promised to myself. I located his village named RAPAKOT. It was approx 10 – 12 km West of the district town of Syangja across ANDHI KHOLA. After visiting Area of MATTIKHAN, East of Syangja extensively for a week, I decided to visit Rapakot. I crossed ANDHI KHOLA at Jaishri Danda and then passed through the villages named SATUPASAL, NAURE BHANJYANG and ARJUN CHOUNTRA. There were few shops and DARUNG KHOLA flowing adjacent to it. Here I came across a girl from Garhi Cantt in Dehra Dun; she had been married to a person from this village who owned this Shop. Any person belonging to Dehra Dun and visiting interiors of Nepal are always treated with lot of awe, respect and affection. I have experienced this on numerous occasions. I remember once, while passing through a village, I requested a lady to give me some water to drink. She got to know I belonged to Dehra Dun. She not only brought a glass of water but some fresh MAI (LASSI) also. That is the awe and respect the simple rural folk held for even the place name DEHRA DUN. This has resulted from the fact that 39 GTC, 58 GTC, 11 GRRC and Gorkha Boys Company had been located there for over three decades. All soldiers going home after Recruit Training, various other duties to their Training Centre and even Pension Drill, always had good memories of the place and shared among all their family members back home in Nepal.

So far I had been walking along a track in which no climb was involved. It was at Arjun Chauntra that I crossed DARUNG KHOLA and started climbing the hill for Rapakot. This village was almost on top of the ridgeline emanating from a another huge ridgeline coming from North and going Southwards towards PAIYUM District. Rapakot was fairly a big village and most of the path had steps paved with stones. While climbing I could well see two large villages across DARUNG KHOLA, named DARUNG and SIRWARI which were inhabited primarily by BAUNS and CHHETRIS. Further West and almost touching the top of the ridge line were two famous GURUNG villages named RAMJA and CHITRE. It was two hours of steep climb before I could reach the village RAPAKOT and located the house of Jabar Bahadur. I could meet his old mother only as father had yet not returned from the fields. His was a modest house containing very basic essentials only. I could well imagine how much difference the presence of Jabar Bahadur would have made in the financial well being of the family. I paid my respects to his mother and introduced myself as a friend of her late son and come to meet her from the Battalion. She obviously felt so very happy. Then in very few words I told her I owed her son an amount of Rs 300/- which I had borrowed from her son, could not return the amount to him earlier when he was alive and now want to pay that amount to his mother. She could not believe it and appeared hesitant in accepting. It was after lot of persuasion that she ultimately agreed to my request and accepted the money.

What she said while accepting the amount, forever reverberates in my mind. She said, "These days people do not return money when the person is alive and you have come from such a far distant place to return the money borrowed from my son who is no more. May God bless you always". It was a very satisfying visit to the house of Late Rifleman Jabar Bahadur Gurung. I fulfilled the promise made to myself and thus God gave me an opportunity to pay my respects to the old lady and in doing so earned her blessings. There could be no better gift than that. I very fondly cherish my few minutes of interaction with the respected mother of my deceased friend, JABAR BAHADUR GURUNG, from the hills of Nepal.

Sun was setting and it was time for me to return, preferably, to Syangja, by the fall of the night. It was already dark when I reached Arjun Chauntra. I wanted to hit the Road POKHARA – BUTWAL so as to catch a vehicle going to Syangja the same evening. However, I could not find any vehicle for quite some time and walked for approx three km along the road during the night and all alone to reach Syangja. Great visit. So thankful to God Almighty to make it possible for me.

# MY EXPERIENCES OF INDO-PAK WAR: 1971

I was commissioned in Army on 06 September 1970 as a 2/Lieutenant and joined First Battalion of the 5th Gorkha Rifles ( FF). This Battalion was to be my home for next 21 years till I moved out to command a different Battalion of the same Regiment in 1991.

A 2/Lt belongs to a special, if not dangerous, category of persons in the Army. He is quite ignorant of the 'Matters Military' but always considers himself to be a storehouse of knowledge. One has to just prod him and then be prepared to listen to the monologue on higher directions of war which he supposed to have learnt as a Cadet in the Training Academy or during Young Officers' Course.

I was a thin, short statured person, weighing just about 48 kg of weight when I joined my Battalion. I remember meeting one Major Malhotra from 4/1 GR who happened to visit our Battalion at a place called Sarol Camp in J&K in October 1970. He first looked at my face and then at a Single Star on epaulettes on each of my shoulders and said, "R V, if one does not notice a Single Star on your shoulders, you can get away as a NCC Cadet". That is exactly what I looked like, more of a NCC Cadet and less of a 2/Lt. Fortunately, I was not alone as a 2/Lt in the Battalion; there were three more of my creed. Those were the days when there used to be a large number of 2/Lts in each Infantry Battalion and Second-in-Command always had a harrowing time controlling his herd.

My first Commanding Officer was Lt Col S C Gupta who later retired as a Major General. He was a thoroughbred professional, fatherly figure and a very fine gentleman of 'Old Order'. Battalion moved from J&K to Ferozepore in Punjab in December 1970. Meanwhile, I proceeded to Infantry School, Mhow (MP) to attend a five months long Young Officers Course. It consisted of Platoon

Weapons Course, Commando Course and Platoon Tactics Course and formed the basic background for a 2/Lt to effectively lead a Rifle Platoon in War. Under the able leadership of our Commanding Officer, we were fully equipped, trained and motivated to undertake any task as assigned.

War on the Western Front commenced on the evening of 03 December 1971 and we were launched into attack on the night of 05/06 December 1971. We experienced very heavy enemy Artillery shelling on the two intervening nights of 03/04 and 04/05 December 1971.

I am writing down few memorable incidents which I underwent personally, was a direct participant and cherish those memories most dearly. There were many more and a small booklet can be filled with their narration.

## ONE

I was No 7 Platoon Commander in 'C' Company of my Battalion. War broke out on 03 December 1971 and our 'C' Company got deployed in Area of Village Rathoke, East of Khemkaran in Punjab. My Platoon was deployed further ahead, South of village Rathoke, facing Pakistani village Sehjra, all along on an escarpment overlooking flood pan of River Satluj. This escarpment was 20 to 30 feet higher than the Satluj River flood pan and the river flowed just few km down South. There were some small rivulets flowing close to the escarpment and East of village Sehjra. My Platoon had dug 10-12 trenches all along the escarpment, all without any overhead protection, primarily to keep effective surveillance and deny any mischief by the enemy.

Around 6 pm on 03 December, the enemy started firing with Artillery and Mortars on our temporary dug-outs along escarpment. Enemy troops were deployed just 600 yards South of us. All our trenches would have been already registered as targets by the enemy Artillery and Infantry Mortars and we experienced fairly effective enemy fire. I had returned from Young Officers Course just few months ago and, like any other Young Officer, was full of

JOSH and brimming with the so called basic knowledge gained at Infantry School not so long ago. I well remembered the advice given by my Syndicate DS of Platoon Tactics, who himself was a War Veteran of India-China War,1962 and Indo-Pak War, 1965. Later, he fought in Indo–Pak War, 1971 as a Company Commander, wounded and even lost an eye. 'While in defence, a Platoon Commander MUST crawl and move from trench to trench in his Platoon, come what may, and motivate his troops to hold on at all costs' was his advice which I followed in true sense. I kept crawling from trench to trench during the time when intense Artillery and Mortar firing was pounding my Platoon defences. In doing that I could first hear the sound of numerous shells leaving the barrels of the enemy Artillery Guns deployed in depth areas, hear the hissing sound of those shells going past my Platoon Location for our depth locations or landing in my locality itself. It continued for the nights of 03/04 and 04/05 December 1971. I must have made numerous round of my Platoon Locality crawling from trench to trench and could well recognise which enemy Artillery shell is going for our depth localities and which was to land on my Platoon locality. Many times you play deadly games in War, this was one for a 2/Lt having hardly a year of Service.

Many times shells landed very close to me during both the nights, yet nothing happened to me or any of my Platoon-mates. However, I noticed both my ears were paining due to the sound and percussion effect of shells landing so close by and for two consecutive nights. I did not pay attention and let go of it. Later, while undergoing various Courses of Instructions in Army Schools, I realised I could not properly hear the instructions imparted by the Instructor standing at the Podium. Many times I used to literally pull my ears so as enable myself to pick up what is exactly was being taught. However, I never reported this to any one owing to my sheer inhibition and ignorance. While undergoing a promotion medical examination from Lt Col to Colonel, 20 years later at Base Hospital, Tezpur in January 1992, I was diagnosed with an old hearing disability. ENT Specialist insisted on downgrading me to lower medical classification. However, on much insistence, he accepted my request for not downgrading. My sole worry was 'what will my Battalion Soldiers think of their medically downgraded Commanding Officer'. I happily returned my Battalion in SHAPE1 Medical Category. Thought of suffering any monetary loss in pension emoluments subsequently for not accepting this hearing disability, was nowhere in my mind. A few people even manipulate some physical disability just prior to retirement for some financial gains.

## TWO

Our 'C' Company was tasked to lead Battalion advance from Village Rathoke, own side of the border, establish Firm Base in enemy territory, secure FUP for Battalion in Phase 1 and then subsequently to be launched in the last Phase of Battalion attack. Firm base was located approx 2 km inside enemy territory and FUP was further 1 km West of Firm Base. FUP and Firm Base were to be secured by a Platoon each from 'C' Company. My Platoon was part of the Troops deployed in securing of Firm Base. My No 7 Platoon along with Intelligence Section were leading the advance of the Battalion from Village Rathoke to Firm Base. We were followed by Company HQ, No 8 and No 9 Platoon, in that order of march. Remainder Fighting Group followed closely thereafter.

We all walked down to Satluj River bed. There was 10 to 12 ft high Sarkanda grass all along and not more than 2 or 3 men walking ahead could be seen by the men following them. Little later, I noticed that the speed of men walking ahead of my Platoon has increased all of sudden and I could then notice just two men walking ahead of me. I moved ahead and found no trace of any person ahead nor could I hear any sound of men walking ahead of me. Tall Sarkanda grass was denying any visibility ahead. Very obviously, my Platoon's link with the Intelligence Section leading the advance to Firm Base had been broken. At that moment I did not know the exact location of Firm Base nor did I have any wherewithal to lead the advance of Battalion to Firm Base at my own. Battalion was on its way to attack an Objective, physical communication broken down, electronic communication cannot be switched on for security reasons! It was my Platoon's responsibility to remain in contact with the leading elements. All these thoughts flashed in my mind in fraction of a second. There was no time to think but to take a quick decision and proceed. Immediately, I passed a verbal message to the elements following me to run behind me fast and asked for this message to be passed to all elements following behind. I literally ran in the straight line for 3 – 4 minutes with prayers on my lips exhorting all possible Gods and Goddesses for establishing my link with the Intelligence Section moving ahead of me. All my Platoon soldiers ran behind me. Fortunately, we were able to link up with the Intelligence Section moving ahead. My happiness knew no bounds. Thereafter, till we reached Firm Base, I ensured link with the leading elements is not broken.

Those 5 - 6 minutes of having broken contact with the leading elements in a Battalion attack seemed to be never-ending for me. The happiness on linking up with the Troops ahead was equally great.

## THREE

Once we reached Firm Base, 'B' Company straight moved to Battalion FUP without deploying in Firm Base and was launched in Phase 1 of Battalion attack. 'A' Company had reached Firm Base, deployed and ready for launch in Battalion Phase 2. I could see every person of 'A' Company leaving Firm Base one after other for their assigned tasks. Suddenly I noticed a person who came close to me and stood in 'ATTENTION' as mark of respect to me. I recognised him immediately; he was Nb Sub Ganesh Bahadur Gurung who till last week was MT Hav, promoted as Nb Sub very recently and posted to 'A' Company as Platoon Commander. That was last I saw of Ganesh Bahadur Saheb as after 2 hours, he fell to the bullets of enemy MMG fire while crossing the minefield in Phase 2 of the Battalion attack. His dead body was recovered from minefield next day. Such is the journey called life during an Infantry assault.

## FOUR

Two Platoons of 'A' Company somehow lost their way and lied

extended between FUP and 'B' Company Objective. Only one of its Platoon could reach and captured its objective. Time was running short. 'D' Company which was reserve to 'A' Company had already left Firm Base for Battalion FUP to be in position for further operations. At this stage CO took a decision of not launching 'D' Company which was reserve to 'A' Company for capturing remainder objective of 'A' Company. Instead, he ordered 'C' Company less a Platoon ( No 9 Platoon was still at the Battalion FUP, having secured it ) to uplift from Battalion Firm Base and come straight to 'A' Company objective which had been partially cleared by its No 1 Platoon and not circuitous route of coming via FUP. L/Nk Dharam Singh Thapa of Battalion Intelligence Section led the C' ' advance and was followed by myself and then remainder Company. We reached 'A' Company objective area well in time to assist it in clearing of remainder portion.

Next day morning it was revealed that, while taking a short cut from Firm Base to objective, we had walked through an Anti-personnel Minefield laid by the enemy. Fortunately, we suffered no casualties. Yes, we did wade through a thigh level deep water stream of River Satluj, East of Sehjra, en route to objective.

## FIVE

When 'C' Company was passing through 'B' Company, it was then in the process of carrying out reorganisation after capture of its Objective. I noticed some of 'B' Company persons were busy digging their trenches with Mess Tins as they had not carried their digging tools with them. I had once heard of likelihood of such an eventuality occurring in war while attending Young Officers Course a Infantry School, Mhow. Lo and behold ! I was fortunate to witness it happening in front of me and so soon in my Army Career.

While passing through Battalion Phase 1 Objective, I saw dead body of a stoutly built Gorkha soldier who had died in Phase 1 of the Battalion attack. I could well recognise him since it was moon light by then. He was Rifleman Om Bahadur of 'B' Company. Within an year of his death, I was destined to visit his house in

Village named PELKACHOUR in Syangja District of West Nepal in October 1972. Late Rifleman Om Bahadur had just been married. I met his 17 years old widow and handed over some financial assistance as given to me by the Battalion. It was very difficult to stand in front of a grieving young widow, parents and other relatives of the deceased. It was after 43 years in May 2015 that I again met this widow at Ranchi when I had gone to attend Raising Day Celebrations of the Battalion. By then she was a 60 years old Lady. I am grateful to her that she immediately recognised me even after such a long time. She burst out her emotions and said, "HUZOOR, TAPAHI LE HAMRO GHAR MA PUGNE KO SAMAY MA MO DHERAI SIYANO THIO" (Sir, when you had visited our home in Nepal long back, I was very small). Indeed, a 17 years old was too young to lose her husband. She never remarried and the Battalion regularly invites and makes arrangements for her visit the Battalion which she often comes. She is a great lady. May God bless her.

## SIX

'C' Company had arrived at the objective of 'A' Company and assisting in capture of remainder objective where enemy elements were still resisting. I was standing close to CO's Party, so were three more officers. We all officers were standing in the open. CO was giving orders to 'D' Company Cdr, on his Radio Set. It was almost bright moon lit night. Enemy MMG was regularly firing at us. One could well hear the enemy MMG Detachment abusing us in chaste Punjabi language. A Soldier from 'A' Company had taken position just a meter on to my Right hand side. He was wearing a steel helmet. Suddenly a small arms bullet hit on his forehead just below the helmet. I heard a feeble 'AAH' sound coming from the soldier and that too once only. He collapsed sideways and died instantaneously. Something fell on my shirt and trouser. It was the blood of the deceased soldier that had splattered over my clothes which I came to know next day morning.

## SEVEN

During Phase 1 of the Battalion attack, a bullet had hit the neck of Rifleman Nar Bahadur Pun, 'B' Company. His bronchial chord had been damaged and he was finding difficult to breathe properly. Since no RMO had been posted to the Battalion, Captain SM Naik had been loaned to us for providing medical cover during the Battalion attack. He joined us in Concentration Area just two days prior to the attack. He made a hole in the bronchial chord just below the wound suffered by the soldier while the Battalion attack was still progressing. Fortunately, casualty survived till his evacuation to Hospital next day morning. A praiseworthy and timely action. This action earned a gallantry award of a Sena Medal to the Medical Officer and thus our Medical Officer Capt S M Naik proudly became Captain S M Naik, SM.

## EIGHT

Sehjra, our Battalion Objective, was a big, old and prominent village in Kasur Tehsil of Lahore District. Its prominence can be judged from the fact that even in the days prior to Partition in 1947, it had a Medical Health Centre manned by a MBBS Doctor. This was verified from Group photo of village elders found in a house which gave these details of a person featuring in the photo. Proportionate to the size of the village and its large population, the village had an equally big Graveyard to bury its dead. This was located to its North- West and not far from the village.

After capture of our Company Objective, our Company firmed in area Graveyard as enemy Air Force and Artillery fire, especially air bursts shells, were expected to cause maximum damage any time at the day break. We started digging trenches to protect ourselves and encountered a strange problem. We could easily dig up to the depth of two to two and half feet and then suddenly come across an empty pit of size approx 3ft by 8 ft. This we found almost everywhere we dug. Day light was close and unless we were properly dug down, we may have to suffer avoidable casualties. Nor, was it possible to shift our area of digging down at this moment as enemy counter attack was expected any time soon. We decided to stay on in the same area what so ever it was.

Next day morning we realised that entire Company had dug down in village Graveyard. One could see the complete skeleton lying in the grave reminding you of the Chart of Human Body seen in the Zoology department in College days; only difference was that part of grave was now albeit temporarily occupied by a living person to safeguard himself from enemy Artillery and Air Burst shells. We prayed for the SADGATI of the departed souls whose skeletons were lying just close by in the grave, may be for hundreds of years and now disturbed by War. I spent few hours of cold December night in one such trench-cum-grave, drenched in Satluj water up to thigh level while crossing the water creak en route to objective not long ago. The sight of a full body skeleton lying so close by for few hours in night, is as fresh in my mind even today and reminds me of my true worth.

# A LIEUTENANT IN TROUBLE AT LUDHIANA RAILWAY STATION

It happened towards the end year 1972. The battalion was located at Ferozepore and a vacancy for detailing an Officer on Winter Warfare Course was allotted. The course was for the duration of 3 months and to be conducted at High Altitude Warfare School, Gulmarg (J&K) from 01st week January to 1st week April 1974. I was overjoyed with the thoughts of spending three months in Gulmarg. I was yet to be posted to any High Altitude Area and my encounter with snow restricted to few inches of it at Mall Road, Mussoorie only. I had all my education in Dehra Dun in 1960s and often used to go to see the first snowfall of the Winter season. Gulmarg in any case was to be much different with heavy snowfall. Also, one attained some level of expertise in Skiing too while undergoing the Course. Due to certain military reasons, all attendees were required to bring their Army issued personal weapon and ammunition from their respective Battalions. Also, it was a rare Course on which Officers were required to bring their Sahayak with them who was also to carry his weapon and ammunition. Course was sort of an adventure for me and suited my nature. Those days trains used to terminate at Pathankot and subsequent journey to Jammu and beyond was to be done by road only. Laying of Railway line from Pathankot to Jammu was almost complete and train services scheduled to commence from Jammu from mid January 1973 onwards.

The course was fun and a big adventure for we all young officers and everyone enjoyed. I did not realise how soon three months passed and found us sitting at the newly opened Jammu Railway Station for return journey to our respective Battalions. I wanted to be rather secure and comfortable in safe custody of my Carbine Machine Gun and ammunition and put them inside my personal

Iron Box which had my name, 2/Lt R V Singh, 1/5 GR, prominently displayed on it. I went a step further ahead and even put the Rifle and ammunition of my Sahayak also in my personal box itself. Owing to its length, his Rifle could not be fitted in my box and was especially disassembled for this purpose. My rail reservation had been arranged in a train going towards Lucknow. There were no AC Coaches those days and Officers were required to travel in 1st Class coach. I was allotted a Lower Berth in a two berth Coupe. Upper Berth was still lying unoccupied. Me and my Sahayak were to detrain at Jalandhar Railway Station which was scheduled to come sometime after midnight. Just before the train started moving, I noticed a passenger entering my Coupe and occupied the Upper Berth. We had a short talk before going to sleep. I learnt he was a JCO from 11 GUARDS Battalion located at Jammu. He was proceeding on annual leave to his home town somewhere in Kumaon and would detrain at Rampur in Uttar Pradesh sometime next day.

I was woken up by the loud cries of Coolies announcing the arrival of the train at Ludhiana Station. I sprang up from my berth in a hurry as I had overshot my scheduled detraining Railway Station Jalandhar by cool 2 hours behind. Also, the train had arrived at Ludhiana Railway Station 5 – 7 minutes ago. I was barely able to bring down my baggage and the train started moving slowly. I asked a Coolie to carry my baggage to the Waiting Room. While assisting the Coolie in placing my box on his head, my eyes fell on the name written on it as Subedar.……….. In a hurry I had brought down the box of my co-passenger and left my box in the Coupe. I soon reacted to collect my box but saw even the last coach of the train going past the Platform. My Sahayak got down at Jalandhar Railway Station and searched for me till the train departed. Since he could not find me at the Platform, he left for the Battalion at Ferozepore.

I was not much concerned about my personal belongings which were in the Box. My worry was the safety of 9 mm Carbine Machine, 7.62mm Rifle and the ammunition kept inside. I stood for a minute, thought over how to proceed further and soon decided to go to the Duty Station Master. He was a very positive person and understood the gravity of the situation. He tried to stop the train at very next Station i.e. DHURI but could not as it had just crossed it. He soon spoke to the Station Master at Ambala Cantt and put me through to the MCO Staff on Railway telephone line. Sub Saheb was contacted, told as to what has happened and what is their inside the interchanged box. He was a matured soldier and understood the requirement and plight of a 2/Lt. I requested him to wait at Ambala Cantt Railway Station but he insisted he will come to Ludhiana Railway Station himself and exchange the boxes and he did that. In doing so, he lost a day of his leave, rail reservation beyond Ambala Cantt and inconvenience faced in doing so. I thanked him wholeheartedly and asked him to at least accept the rail fare incurred on his to and fro journey; he most respectfully declined that. I am ever grateful to the JCO from 11 GUARDS. He was very cooperative, considerate and thoughtful. His actions were, indeed, praiseworthy. May God bless him.

# A PERSON WHO DREW PENSION FOR HALF A CENTURY

In 1972, I first time visited Regimental Recruiting Areas in Nepal. I took two months of annual leave from mid-September to mid-November and went on a long trek to Western Nepal to meet retired soldiers and dependents of my Regiment. The itinerary of my visit had been planned in a manner that I am able to meet maximum number of serving and retired soldiers in their villages. Area approx 150 to 200 km around Pokhara has quite a dense population of retired and serving Gorkha soldiers and is an ideal place for interaction with regimental fraternity.

The area of BHOKSING – PAIYUM is approx 50 – 60 km South of Pokhara and 15 km West of road leading from Pokhara to Butwal in the South. This area is mainly inhabited by Gurungs and 5 Gorkha Rifles has a large recruitment base in this area. It took me three days of comfortable walk from Pokhara to reach BHOKSING and then to the house of Nb Sub Nar Bahadur Gurung, who later on rose to become Sub Maj (Hony/Capt) and ADC to The President of India. Nar Bahadur Saheb belonged to a truly illustrious family. His grandfather was from my Battalion and retired as Sub Maj / Hony Capt, awarded OBI, SARDAR BAHADUR, IDSM prior to First World War. His father, Jemadar Dhan Kaji Gurung was also my Battalion and died in the battle of Cassino, Italy, during Second World War and was awarded Military Cross. His grandfather had constructed a huge house in his village, deep in the interior of Nepal which resembled a present-day Flag Staff House in a Military Cantt. His mother had been providing free accommodation and food in her house, school fees and other allied expenses to 7 – 8 bright students of the area around. All these students were from financially poor background and admitted in the nearby High School at BHOKSING. Nb / Sub Nar Bahadur Saheb was born two months after the death of his father during 2$^{nd}$ World War and primarily brought up by his dotting grandmother.

As a word has already gone around about my intended visit, number of old, some of them were very old retired soldiers of my Battalion / Regt had come to meet. All of them were duly attired and almost all JCOs were wearing respective Regimental Blazers. Some came wearing medals awarded for the services rendered in the First and Second World Wars and even the 1948 and 1965 Indo-Pak Wars. What a proud moment for me to watch them so proudly attired! Each one of them exhibited a great sense of belonging to the Battalion and Regiment.

Sub Indra Prasad Gurung of the Battalion belonged to a nearby village named Dune Danra. His was again a very illustrious family with four generations having served in the Battalion beginning around 1880 onwards. It was here at the house of Sub Indra Prasad Gurung that I met Sub Dharam Singh Gurung who retired from the Battalion in 1937 and had participated in First World War all over Middle East and part of Europe with the Battalion and seen the Great War from a very close distance and were so keen to keen to know the performance of the Battalion in recently concluded Indo-Pak War of 1971. He sat with me, a mere 2/Lt, for full 2 hours on that day, came back again in the evening and had further discussion with me. A great soldier, very proud of having served in the Battalion and Indian Army, which has acknowledged his services and provided them pension for such a long time.

It was in Paiyum Bachha I met Hony /Lt Uman Singh Gurung, who had retired from Service in the Battalion in 1970. He had accompanied 2/Lt Harpal Singh on a Long Range Patrol as a Havildar when the Battalion was deployed in SIANG-SUBANSIRI Area of now Arunachal Pradesh in 1958-59. 2/Lt Harpal Singh slipped into the fast flowing SIANG / Subansiri River and drowned unnoticed as he was downstream of the remainder Patrol Party. Even his dead body could not be traced at that time. However, I happened to meet another JCO at the house of Sub Indra Prasad Gurung in Paiyum area. He was from 9 ASSAM RIFLES which had occupied a Post in the near vicinity during those days and gave an additional input on this unfortunate incident. He stated that the troops of his Post had noticed a dead body of some Army person wearing Gorkha Regiment Officer rank epaulettes on his shoulders, stuck on the River bend for some time. Apparently, the dead body was of 2/Lt Harpal Singh, though this information could never reach the Battalion. Such is the journey called life. May God grant SADGATI to the departed Soul.

Enroute between the village of Sub Indra Prasad Gurung and Nb/Sub Nar Bahadur Gurung fell the village of Sub Bir Bahadur Gurung. He was from 1/5 GR and retired from Service in 1970. He must have been an excellent JCO in his time. I met the father of Sub Bir Bahadur Gurung who was around 85 years of age, served with the Battalion for 15 years and retired as Rifleman in 1922, half a century ago. He fought the entire First World War as part of the Battalion and participated in all operations fought by the Battalion in MESOPOTAMIA, now Iraq, Italy and many other European Countries. He was so happy to have earned his pension since last 50 years and grateful to Indian Government and the Battalion. It was truly a long journey to earn pension for 50 long years and not very many of we soldiers are endowed with such a blessing by the God Almighty. His grandson, named Ram Bahadur Gurung, was undergoing recruit training at 58 GTC at Dehra Dun and intending to join 1/5 GR. It was great meeting the father-son duo and their Regimental espirit. I truly felt happy and proud of meeting both of them.

Strange are the ways of destiny. His grandson, Ram Bahadur Gurung, son of Sub Bir Bahadur Gurung, instead of joining 1/5 GR joined 4/5 GR. Later, after almost 20 years of this incident, I took over command of 4/5 GR, found him as a JCO in the same Battalion and doing as Nb Sub / Adjt. Our relationship went much beyond CO and Nb Sub / Adjt level; our trust and faith in each other was always much deeper than the formal relationship. He was an excellent Football player and retired as Sub Maj of the Battalion in due course of time. An illustrious family, I am indeed proud of having met them at their home deep in the interiors of Nepal. May God bless them wherever they are.

# BEWILDERED BISHNU BAHADUR

I joined Battalion from Officers' Training School, Madras in September 1970 and so did Rifleman Bishnu Bahadur Pun from 58 Gorkha Training Centre, Dehra Dun. A simple, shy, stoutly built and bit dark complexioned for a Gorkha. He belonged to a village named CHAURAKHANI in Myangdi District of Dhaulagiri Anchal in Western Nepal. PUN is a sub division of MAGAR, which generally inhabit middle reaches of Central and entire Western Nepal right up to Indo-Nepal Border. There is a further sub-division of PUN who resides in the upper reaches of Dhaulagiri Anchal and are known as KHAM. KHAM have their own dialect which is different from MAGAR or even PUN dialect. Bishnu Bahadur Pun's village was located in these higher reaches, almost touching the base of DHAULAGIRI Mountain and last inhabitation on way to this beautiful Mountain. It was difficult to trek around in this wilderness, more so when you are alone and no one in position to keep track of your welfare and whereabouts. You enter the hills, melt into them for the entire duration of your trek and surprise all by coming out one fine day, safe and sound with lot of happiness and contentment writ large on your face.

I undertook my second Trek to Nepal, this time for a period of 3 months, during October to December 1974. Army HQ was kind enough in allowing me to avail year 1971's unavailed annual leave too during this year. I collected details of soldiers who will be on leave during this period and, to an extent, planned my itinerary accordingly. I could not visit Myangdi district during my previous trek undertaken in 1972 and was definite to visit the area this time. I met Bishnu Bahadur Pun in first week October 1974 when he was proceeding on his four months long annual leave. I was to follow a week later and told him of my likely visit to his village in Nov or Dec. He laughed and told me no one from outside has ever visited his village as it is located in a very remote area.

I took a Bus from Pokhara to Palpa. It was an old vehicle and mid-way I had to change the vehicle and hitch-hike into a private Load Carrier vehicle. We were three persons sitting in the rear of the vehicle, one of them was a middle age person in his mid 40s while other a young person in his early 20s. During our talk, I told them about myself being an Officer from Indian Army and presently going on a Trek to Palpa, Gulmi and Baglung. A little later, we had a short halt near a Tea Shop en route and again sat behind the vehicle for our remainder journey. I noticed the young boy was no longer with us. I enquired about the missing person from his friend. His reply made me laugh on the innocence of the missing young boy. He told that his young friend was actually a deserter from Indian Army and he mistook you to have come to arrest him. He got worried and vanished before the vehicle reaches Palpa where Nepal Police has a proper Post.

It was long trek from my last Road Head Tansen to MYANGDI criss-crossing GANDAKI River numerous times. DHAULAGIRI Mountain looked so small though beautiful from ARGHA KHANCHI and its view kept getting bigger with crossing of every district, namely, PALPA, ARGHA KHANCHI, GULMI, BAGLUNG and lastly MYANGDI where one almost reaches the base of this Mountain. GANDAKI River emanate from TIBET and enters Nepal just North of MUSTANG, separating ANNAPURNA and DHAULAGIRI Mountains near JHOMOSOM and traverses through entire Western Nepal from North to South before entering India. I again followed Kali Gandaki River from Baglung to Beni which was then a small town and presently HQ of Myangdi District. I now followed Myangdi River and reached a village called DARBANG where Myangdi River coming from North takes a sharp Eastward turn to join Kali Gandaki at BENI. Next day I was to reach a village called BIMA which was half a day's march further North and on the East of Myangdi River. Bima is the village of Nb Sub (later Hony Capt ) Min Bahadur Pun of our Battalion. He was Rcl Gun Det Commander with our Company during 1971 Indo-Pak War and I had lot of regards for him. He had already sent a message to his family and other Ex-servicemen of the village of my likely visit. I spent one full day at his house. Nb Sub Min Bahadur Pun Saheb was in the Battalion and his young son, who later joined British Gorkhas, became my friend and guide for the stay in their village. Numerous Ex-servicemen from my Battalion/Regt and other Gorkha Regiments joined up in the evening. It was an excellent meeting with all the BURHOS and many reminiscences of their long service while with Indian Army were revisited. Few of them were expectedly high on drinks. There was one Subedar ( retd ) Manu Pun from 2/3 GR who was truly glad with his local brew and wanted to verify my identity. He asked me to produce my Identity Card. Fortunately, despite instructions from the Indian Embassy in Kathmandu to the contrary, I had carried my Identity Card. He almost snatched that from my hands and started scrutinizing the details with his fading eye sight. After half a minute he said, "You said you were commissioned in 1970 but here it is mentioned as 1947". Actually he read my date of birth and mistook it as date of commission. He raised few more interesting queries. I had to simply continue answering his queries. This session lasted for quite some time and other ESM gathered were feeling uncomfortable and apologetic. Nothing much could be done to a person otherwise high on drinks. Little later, he respectfully handed over the Identity Card thus ending my interesting ordeal. Next day was my departure for the village of Bishnu Bahadur Pun which lied further 2 days march ahead in the barren, high mountains with sparse population. All those ESM who were with me last evening had gathered around to bid farewell. Subedar Manu Pun Saheb was there in the front. He very sincerely apologized for the inconvenience caused to me in the previous evening and was almost in tears. He further added,"Saheb, it is our officers who advise us during our retirement drill in the Centre to be careful of spies in Nepal". He was correct to an extent as MAOIST trouble had already started in Nepal in end 1960s and he was making an attempt to verify antecedents of a visitor from a foreign country. On completion of my trek in end-Dec 1974, I met Lt Col , later Lt Gen, R K Anand, 2/3 GR who was then Military Attache in Kathmandu and happened to be CO of Subedar Manu Pun in the Battalion. I narrated this interesting incident to him. He had a hearty laugh and asked me to give the address of his JCO as he had been trying to locate him for quite some time. Lt Col R K Anand asked me if I had anything more to discuss. I asked him how come many of the JCOs are blessed with children in their old age after their retirement from service. I could see he was barely able to control his laughter at my childlike ignorance. He explained the reasons as prolonged separation from their families in Nepal, non-availability of sufficient married quarters to the JCOs /OR of Infantry Battalions during peace tenure and busy schedule of work in the Battalions. He was so very right.

Back to progress on my trek to the village of Bishnu Bahadur Pun. I had to walk for two more days. Track from BIMA to KUINE KHANI was quite slippery and involved a tough continuous climb of more than three hours. KUINE KHANI was a small village and in total wilderness. Only a few people live here during winters and majority migrate to lower reaches. Shopkeeper where I had tea advised me not to proceed for CHAURAKHANI at this late hour. In any case I was dead tired. I met one Sub Harkaman PUN from 1/4 GR. He also advised me stay put for the night there only. Height of this village was around 10,000 ft and biting cold in the night. It was here that I first heard people talking in KHAM Dialect. It was different from Magar Dialect and closer to GURUNG and THAKALI Dialect. Sub Harkaman Pun again came to meet me before my departure next day morning. He saw me shaving and requested to give my Razor for shaving the head of THAKALI in whose house I had stayed during the previous night. Such basic essential facilities are generally not available in those remote areas and people have to wait and adjust accordingly. My next destination - CHAURAKHANI – was still a day's march ahead. Fortunately, I had company of a Lady who basically hailed from this very village KUINE KHANI. She along with her 10 year old daughter was returning to her in-laws village named PALIKHARKA, a few km ahead and which fell on our route ahead. Her husband was serving in ASSAM RIFLES in Arunachal Pradesh. A resident of her village Kuine Khani returning from another village ahead recognised her and enquired as to when did her 'LAHURE' come on Leave. The Lady blushingly replied that the person accompanying her was not her 'LAHURE' but a stranger from India and visiting next village CHAURAKHANI. We travelled together for 2 -3 hours then she veered off to her destination.

The track was only for name sake and any wrong turn taken may lead to spending the night on some tree or in cold wilderness. Fortunately, I was able to reach my destination, CHAURAKHANI, and was guided by a villager to the house of BISHNU BAHADUR PUN. His father welcomed me at the door. Bishnu Bahadur was inside his house and did not know of my arrival. His father called him out to meet his friend from India. He came out and saw me; it was a sight to see the feelings on his face. He was closing / opening and rubbing his eyes time and again. He simply just could not believe what his eyes were seeing in front of him, Captain R V Singh, in flesh and blood, standing at the door step of his house in this so far off and remote village. How can it be possible!! He was unable to control feelings of happiness in his heart and how to express them to me. He was completely surprised and had not even dreamt of that I really meant to visit his house one day. Surprise and gratitude was so writ large on his face. My dear friend BISHNU BAHADUR PUN WAS TOTALLY BEWILDERED. He had a large family of six brothers and four sisters, all younger to him. He had been married two years ago. I spent two days at his place and Bishnu and his parents went out of their way to look after me. Very rarely any visitors come to their village, hence, the celebrations. Bishnu wanted me to accompany him on a hunting expedition of 2 – 3 days duration in the upper reaches which I politely declined. However, he wanted to shoot a GHURAL, mountain goat, found in higher reaches so as to entertain me with a good non-vegetarian dish. He took his father's old single bore locally made Gun and went alone on a 2-3 hours long outing in the nearby mountains to try his luck. He did come across a Ghural but the old gun refused to cooperate and was unable to fire. Poor Bishnu was disappointed and came back empty handed. A beautiful, fast flowing mountain stream named THULO KHOLA, emanating from a lake at the base of DHAULAGIRI Mountain, was just adjacent to the village.

Markets are quite far and daily necessities, like, salt, sugar, tea, condiments etc either had to be arranged well in advance or substitutes found locally. There was no tea or sugar in the house and natural honey was used for sugar and leaves of a specific plant used in place of Tea leaves whose taste and aroma was equally good, if not better than any original tea leaves. In fact next day when I left the house of Bishnu Bahadur Pun, his mother had prepared puffed rice meshed in honey for my journey ahead. It was very tasty snacks and sufficed me for next three days. It was truly a memorable visit to CHAURAKHANI and two days spent with the family members of BISHNU BAHADUR PUN. May God bless him, his parents and all his family members.

I continued my journey next day to POKHARA which was 5 to 6 days of walk ahead. I wanted to visit MUSTANG BHOT via JHOMOSOM as also pious MUKTINATH. Lot of disturbances had been caused by the MAOISTs in these areas. I had met District Administration at Baglung to obtain permission to visit Muktinath and Mustang. However, for safety reasons, District Administration did not permit me to go beyond a place named DANA along Kali Gandaki River. JHOMOSOM was just half a day's march ahead. MUSTANG and famous Hindu Shrine of Muktinath were not far from Dana and still I could not visit these two places due to the disturbances caused by the Maoists in the area ahead. I took the return journey route from DANA via GHARA, KHIBANG, SIKHA, GHOREPANI, BIRETHANTI, PHEDI, DHAMPUS and lastly POKHARA where from I was to take a transport for my return journey back home via KATHMANDU for necessary debriefing at Indian Embassy.

*My First CO, Maj Gen S C Gupta, VrC and Sub Maj G B Rana, IOM.*

*Sub Maj N B Gurung, ADC to the President.*

*Sub Maj A S Ale, ADC to the President.*

*Sub Maj Indra Prasad Gurung.*

*With Company Senior JCO Sub Dharam Singh Thapa, 1984.*

*With 'Bewildered Bishnu Bahadur' in Nepal,1974.*

*In Memory of Two Friends Col D J and Maj Gen Kochekkan*

*At SAMI TOP, Arunachal Pradesh, 1985.*

# A LOVE STORY WENT SOUR

Some people impress you in the very first meeting you have with them. My first meeting with Havildar Kul Bahadur Thapa went something like that. He was a smart person though slightly dark complexioned for a Gorkha. He was Intermediate passed, quite a high educational qualification for a Gorkha Soldier of 1960s when even illiterates could be enrolled. He did very well on the Pioneers Course in College of Military Engineering, Pune and had been appointed Pioneer Platoon Havildar of the Battalion.

I first interacted with him in mid-1972 when Battalion was at Ferozepore. I was a 2/Lt and detailed as Weekly Duty Officer and required to carry out a host of checks in the Battalion by day as well as by night. I was taking a night round of the Battalion Lines and, around 11.30 pm, noticed some light still switched on in a small room in contravention to the laid down instructions on the subject. Self, accompanied by Duty JCO, visited the Room and found a person reading an English language Newspaper 'The Junior Statesman'. Later I learnt that he had personally subscribed for the Newspaper on regular basis. This was not a common incident and left an impression in my mind this person is different from others. Next, I interacted with him was an year later in June/ July 1973 when he accompanied me as part of Battalion's Advance Party to HAYULIANG in Lohit Division of Arunachal Pradesh and did a commendable job as Pioneer Platoon Havildar.

Sometime in early 1974, he was appointed as Officers' Mess Havildar. It was an important appointment for a Senior NCO and generally given to prospective future JCOs. Most of the Officers used to break off from the Mess around 10 pm after having dinner, yet there were 2 – 3 Officers who would continue to enjoy their drinks for some more time. It was during this extended stay of few officers in the Officers' Mess after dinner that an unfortunate incident happened which plummeted the fate of Hav Kul Bahadur Thapa.

It was a usual day. Most of us had departed from the Officers Mess around 10 pm after dinner; three Majors were still there enjoying their drinks. After a while, Mess Havildar also left after giving certain instruction to the Wine NCO to take care of remainder Officers. Little later, a specific brand of Whisky being enjoyed by the officers finished and required recouping from the Tent of Mess Havildar. Mess Havildar could not be contacted on local telephone and Wine NCO showed reluctance in going to the Tent of Mess Havildar. Ultimately one Major went and called Mess Havildar outside. Finding no response he barged into the Tent. What Major Saheb saw inside the Tent left him aghast. Hav Kul Bahadur Thapa was trying to push a young local MISHMI girl under his makeshift cot and both of them were almost naked. This was a serious violation of discipline by Havildar Kul Bahadur Thapa. He was duly punished for the offence and sent to 58 GTC at Dehra Dun for premature Pension Drill. Since Nearest Railway Station TINSUKIA was two days drive further down, an escort Party took him all the way and dropped him at Railway Station.

We thought that the case had been finally closed but it was not to be. After 8 – 10 days we received a telephone call from nearby Police Station at TOILONG asking us to collect a soldier of our Battalion, arrested by them from a nearby village the previous night. He was apprehended by the villagers while intruding in a house to kidnap a young girl and claimed to be from our Battalion. We collected the intern from the Police custody and brought him to our Unit Lines. Surprisingly, he was none other than Havildar Kul Bahadur Thapa who had been dropped at Tinsukia Railway Station 10 day ago.

What actually had happened was that Havildar Kul Bahadur Thapa, though much married and two young kids behind at a village in Palpa district of Western Nepal, fell madly in love with the young MISHMI girl and so was the girl. Once dropped at Railway Station by the Escort Party, he never boarded the train for Dehra Dun but returned to HAYULIANG to take her away to Nepal. Their plans misfired as he got caught by the parents of the girl and handed over to the Police. This time Battalion took no chance of just leaving him at Tinsukia Railway Station but the Escort Party took him all the way to 58 GTC, Dehra Dun for completion of his Pension drill.

Alas! A bright career was cut short as he could not control his emotions. I as a person always felt sad for him as he had lot of potential to do well otherwise. May God grant him SADBUDDHI to him to live a happy life wherever he is.

Not to be left behind, another story was soon in the making. A senior Major, having a happy married life back home, got emotionally involved with a Civilian Lady in the nearby town of TOILONG. It was surely a one-sided affair. Yet Major Saheb surely was head over heels in this affair, come what may. That gave the young officers something to chat about the interesting situation likely to unfold soon unless there was something divine intervention to prevent.

Yes, it was a divine intervention. Before it could reach any damaging proportions, Army HQ came to our rescue and unfortunate drama got averted before it could be staged. Major Saheb got posted out at a short notice and soon he left the Battalion. I do not know how the Senior Officers reacted but we the young officers did heave a sigh of big relief, though remained little apprehensive for long. Whenever there was an incoming call from TOILONG Police Station, we used to worry - KAHIN YEH WOH MAJOR SAHEB TOH NAHIN - KAHIN YEH WHO MAJOR SAHEB TOH NAHIN. Later, Battalion got de-inducted from Arunachal Pradesh to Mizoram and the Chapter was finally closed.

# VISIT TO LAMJUNG DISTRICT IN NEPAL

LAMJUNG, the place name generates a host of feeling in one's mind. Cast off your hearing faculty for a while and listen to the sound made by this word L A M J U N G again, this time listen through your heart. It makes so sweet a resonance. LAMJUNG, the word seems to be seeped in antiquity of time period and binds it so beautifully well with BHOT, other name of the region bordering TIBET on the North and Nepal on the South. Cool resonance of this word is so different from names of other districts of Western Nepal. Notice the masculinity of the District name 'GORKHA' or femininity of the District names 'KASKI' or 'GULMI' or simplicity enshrined in District names 'PALPA', 'ROLPA', 'DOLPA' , 'TANHU' and even '4000 PARBAT'. Word LAMJUNG lifts you to its oldness of its inhabitants, an interesting mix of their version of Hinduism and even culture and traditions with those enshrined in BUDDHISM. Inhabitants of higher reaches of the Himalayas often brought Rock Salt from BHOT, to say, Tibet, before Chinese occupied it over seven decades ago. Caravan of people from numerous villages used to gather and go to Tibet to bring back their annual requirement of rock salt.

LAMJUNG District is primarily inhabited by GURUNGS alone who possess a different dialect, named GURUNG KURA, unintelligible to other groups amongst Gorkhas. Though Gurungs are found in many other districts too and in large numbers, eg, 4000 PARBAT, KASKI, PAIYUM, yet their concentration is huge in LAMJUNG district. I visited Lamjung district first time during my trek in 1972 and could barely touch the Southern reaches of the district where it meets Tanhu district on its South. MIRLUNGKOT , a huge ridge line extending from MARSYANGDI River in the East, bordering with Gorkha District to Kaski District in the West is the dividing line between LAMJUNG District in the North and TANHU District in the South.

I planned to visit almost entire LAMJUNG District during my Trek in 1974. LAMJUNG district occupies comparatively higher reaches of Himalayas. On its North are forever snow clad high mountains, GREAT HIMALAYAS, which have many of the world highest mountains and form border with TIBET. Gorkha District is located on its East, Tanhu on South and Kaski on West. For ease of understanding, LAMJUNG District can be divided in five different parts each having a peculiar name. We start from Pokhara side where LAMJUNG district touches Kaski District on its West and separated by a fast flowing Nala named MADI KHOLA which emanates from the Annapurna Group of mountains in the North. First ridge line flowing from North to South, keeping Madi Khola on its West, is named PANCHGAON as it has five villages on it, namely, Tangting, Ghyamrung, Yangjakot, Warchok and Bhachok. Further East are three smaller ridge lines coming down from high mountains and each having one to two villages, a total of six villages, namely, Naidhar, Rabedanra, Pasgaon, Singdi, Chara and Bhoje. Jointly, this group of villages is called CHHEGAON. Further East, a big ridgeline rolls down from North to South and has ten big Villages located on it and aptly called DASTHAR. Prominent villages on this ridge line are Ghanpokhara, Ghalegaon, Rabase, Majgaon, Baglunpani, Thuloswara, Turlungkot and Nalma.

East of this ridgeline is a big fast flowing stream coming again from Annapurna Group of mountains, named MARSYANGDI. A good track starts from DUMRE on road Pokahra – Kathmandu, goes Northwards all along Marsyangdi right upto Manang Bhot behind Annapurna Group of Mountains, connecting it with famous Hindu shrine of MUKTINATH and then meets Kali Gandaki River near JHOMOSOM. It is a famous trekking route and presently known as ANNAPURNA CIRCUIT in tourism industry. Other half of Annapurna Circuit includes Track going from Pokhara Westwards to Birenthanti, Ulleri, Ghorepani, Ghara, Sikha, Dana, Tukche, Jhomosom, Kagbeni and Muktinath. This trek starting from Pokhara and going to Jhomosom / Muktinath is a more famous trek and was immensely popular among HIPPIES of 1970s era. Even very famous personalities of USA and Europe used to undertake this trek from Pokhara to Jhomosom and then fly back from Jhomosom to Pokhara. In 1972, I came across Mr Robert Macanmara, erstwhile President of World Bank and then Secretary of State, USA Government going on a trek from Pokhara to Jhomosom with barely 5 – 6 persons accompanying him and no pomp and show attached. He was a very senior person in US Government and could still spare time to undertake this Trek in wilderness. Such like hobbies are somehow not very popular among we Indians.

Area East of MARSYANGDI is called BARATHAR and, as the name suggests, it has 12 villages. Beyond this further East and touching Gorkha district, is a huge landmass, numerous ridge lines rolling down from North and some flat area in between and is called RAINAS. Between Barathar and Rainas is another small ridgeline named Lamjung Darbar. During September 1972, I followed this track emanating from Dumre on Kathmandu – Pokhara Road for going to a village on MIRLINGKOT ridge line. TANHU lies on its South and LAMJUNG on its North. It may be of interest that TANHU is predominantly inhabited by MAGARS and LAMJUNG by GURUNGS. En route, I had planned to halt at the place of a retired JCO from my Battalion named Hony Lt Bhagta Bahadur Ale in his village named BARBHANJYANG which was a days march away from DUMRE. BHAGATA BAHADUR ALE Saheb had retired in 1970 . He had a son named Rifleman Ram Bahadur Ale who was serving in 3/8 GR and was representing Services in Football. I spent a night at Ale Saheb's house. I noticed a Group photo, duly framed, and proudly displayed in the verandah of his house. It was Photo of a Recruits Platoon under training at 58 GTC sometime in early 1960s. Ale Saheb, then a Recruit Platoon Commander, was proudly seen sitting on a chair in centre of the front row. What surprised me more was that most of the recruits, as I could then recognize them in the photograph, were representing our then Battalion Football Team. Tika Ram Thapa, Prem Prasad Gurung, Mohan Singh Rawat, Surendra Singh Rawat, Kharak Singh Thapa, Khim Bahadur Thapa, all were there. I am sure Bhgata Bahadur Ale Saheb had ensured all of them get posted to his Battalion.

LAMJUNG district is a big recruitment base for all Gorkha Regiments. One always comes across number of serving and retired soldiers in each village. SIKLES was the last village of KASKI district and from where I planned to enter LAMJUNG district. It is separated by MADI KHOLA which had fast flowing and freezing cold water. Those days there were rarely any bridges provided anywhere on these rivers or Nalas by Nepal Govt. At few places Indian, British or US Governments had launched some steel rope bridges which were far and few. More often, local villagers used to put a log bridge at suitable places as alternate means for crossing the river stream and these were often swept away during rainy season every year. One had to be very careful while crossing over such temporary log bridges or wading through the fast flowing freezing cold water stream, especially when moving all alone. If R V Singh fell down, there will be no trace left of anywhere as I was always on my private trek and moved all alone. I remember an Officer from my own Battalion 1/5 Gorkha Rifles, named 2/Lt Harpal Singh, while on a long range patrol in Arunachal Pradesh way back in 1958-59, slipped down in fast flowing Subansiri / Siang River. Even his dead body could never be traced. I had been trekking in these beautiful, interesting and yet very difficult areas as a private individual and not on any Government sponsored visit. I had to exercise bit of caution on myself every time I made crossing in a mountain stream, be that over a log bridge or just wading through the fast flowing stream all alone.

It took me five days of leisurely trek through these high mountains to cross entire PANCHGAON and CHHEGAON and arrive at the base of DASTHAR. Most of the villages I passed through were either the last or second last inhabited village on that Ridgeline. It was end November 1974; weather had gone quite cold and snowline coming closer day by day. Especially, nights were quite cold and a small sleeping bag carried by me sometimes fell short of beating the cold and an extra blanket or quilt was not always available. En route, I met numerous serving and retired soldiers of my Battalion as well as from numerous other Gorkha Regiments and whose so very fond memories I cherish till date.

First time I heard GHANPOKHARA village on DASTHAR Ridgeline was in mid 1974 when Battalion was located at a place called Hayuliang in LOHIT Sector of Arunachal Pradesh. Brig Jaggi Sikand, 2/5 GR, was commanding Brigade in neighbouring Division and came on a flying visit to carry out some Operational recce. He had been Additional MA at Indian Embassy in Nepal as a Captain sometime in 1950s and had visited this village. Before departing for helipad he spoke to the Mess Havildar customarily and enquired the place he belongs to in Nepal. Answer was GHANPOKHARA and Brig Sikand gave a 5 minutes long talk on beautiful surroundings in which this village was nestled, arduous approach leading to it and fascinating view of entire area on to East, West and South of the village. I was enamoured of what he said about this village and made it a point to visit this village during my next trek which I had planned few months later.

It was already around afternoon when I crossed the stream separating last village of CHHEGAON complex with DASTHAR. GHANPOKHARA was the last and highest village on DASTHAR ridge line. I had to undergo a steep climb of three hours to reach GHANPOKHARA and had to make it before darkness fell. After three hours of continuous climbing I was able to reach a village called Ghalegaon and was told that GHANPOKHARA was still 45 minutes of climb ahead. Residents of Ghalegaon told me that most of the residents of GHANPOKHARA village had already migrated to alternate places to lower heights due to heavy snowfall in their village during winters and I may hardly find any one up there. One of my Battalion Soldier, named L/Nk Budhi Bahadur Gurung belonged to this village and was on annual leave. I planned to visit him and continued to walk with that aim in my mind. It was almost getting dark by the time I reached GHANPOKHARA. Entire Village gave a deserted look as I could see no movement in the village. Fortunately, I saw a person rushing down the path. He was so surprised seeing me there at this odd hour and advised me to enter in any house where I notice some smoke coming out otherwise may not find any place to rest for the night. He soon noticed a house from where some smoke was billowing out. I requested him to help me finding some place in that house for the night. There was an old lady alone in the house. He spoke to the old lady requesting her to permit me to spend the night in the house. He soon left for Ghalegaon.

Old lady, my respected host, was more than 90 years of age. Her entire family, as an annual feature, had migrated few km down slopes where they had their fields for cultivation. The old lady had decided to stay back in the house alone for the entire Winter season. It was a medium size house having slate roof with a central fire place as was generally found in all houses in hills of Nepal. Fire was burning in fire place and some extra firewood also kept beside for use during the night. Once I had put my Rucksack down, Old Lady clarified there was no tea or food to offer. I thanked her for the concern and told her not to worry as I have some biscuits with me which we both shared. She appeared hesitant to talk to me. Initially, she was scared and even said so while mumbling to her own self, "AJU TOH YO PARDESI MAILE MARI DINCHHA HOLA ( This stranger might kill me tonight or words to that effect)". We soon started talking about her health and family details to start with which she readily answered. I found she wanted to talk more and had a good memory for her age. She was married almost 70 to 75 years ago. Journey of her life had been quite eventful. In brief, it went as follows. She was married at an early age when she was barely 17 years old. Her husband was a LAHURE, meaning a soldier of British Indian Army, and was serving in some Gorkha Battalion in ABBOTABAD, now in North West Frontier Province, Pakistan. She accompanied her husband all the way, in those primitive days of transportation, from GHANPOKHARA to ABBOTABAD to be with her husband. She had barely spent two years at ABBOTABAD and lost her husband in some operation launched by her husband's Battalion against Pathans. She was too young, barely 20 years of age and became a widow. She returned to her village. Trasportation system was in primitive stage those days. There was no train facility from Gorakhpur to even Nepal border. Her walk commenced from Gorakhpur itself all the way right up to her village GHANPOKHARA. After some time her parents married her to another person of the village and ever since she had been staying there in the village only.

During my talks with her she made a mention of the word NAKKALE PALTAN and stated her husband was from NAKKALE PALTAN. Hearing this word, I got reminded of having heard this name from Senior JCOs of the Battalion when I was a 2/Lt . In olden days our Battalion was called NAKKALE PALTAN in the interior of Nepal. Those days, Gorkha Soldiers from Nepal, were permitted to accumulate annual leave for three years. While proceeding on 6 to 7 months long accumulated annual leave, Soldiers from this Battalion used to change in to full Drill Order of Army dress before entering Nepal border and carried on walking in Drill Order till they reached their respective homes. This was a strange practice followed by the personnel of one specific Battalion and was generally called in that part of Nepal as NAKKALE PALTAN and its soldiers recognised by their Drill Order dress while walking on mountainous tracks to their villages.

The word NAKKALE PALTAN as mentioned by her during the discussion and having heard of this word from our old JCOs, helped me join the dots. I could correlate that the husband of the old Lady sitting in front of me and talking that day was from own Battalion and whose husband had died fighting on North West Frontier. This must have happened sometime around the turn of 19th Century ie year 1900 or may be 1901, much before even 1st World War. What a grand co-incidence of meeting so old a War Widow of my Battalion in 1974 whose husband had died almost 73 years ago. If she was to be alive today in the year 2021, she would have been almost 140 of age. May God grant SADGATI to her Soul.

Next day I took leave of the grand old lady and thanked her for granting me permission to take shelter in her house as also listen to her talks. I learnt that the soldier from Battalion whom I had come looking for at GHANPOKHARA had also shifted out to a nearby village 3-4 km downhill. Next day, I met him in his that village and spent two days with him. Few more soldiers of our Battalion who hailed from the nearby villages also joined up. Being at quite an elevated place, one could enjoy seeing KHUDI KHOLA, KHUDI BAZAR, MARSYANGDI NADI, BARATHAR and even RAINAS so clear from here. All 10 villages of DASTHAR DANDA and even area of DHULPUR DHODRE further South where from Hony Capt AS BAHADUR GURUNG of our Battalion belonged. He had retired in October 1970 when I had just joined the Battalion on my commissioning. I was fortunate to have visited his village Dhulpur Dhodre and stayed with him for two days during my first trek in October 1972. I took my return journey route by traversing along DASTHAR DANDA, meeting the family members of numerous serving and retired soldiers of the Battalion before entering into KASKI district on my way to POKHARA. While walking down from Dasthar Danda, I had halted at small tea shop en route where a Lahure had halted for the previous night. He was carrying a small earthen MATKI whose head was covered with a red cloth and tightly tied around the neck of the earthenware. Lahure left for his destination before I could enquire from him what he was carrying in that earthenware. I asked THAKALIN, the Lady Shop Owner, as what the Lahure was carrying. She laughed at me and told it was the PHOOL (ASHES ) of some deceased soldier of the Lahure's Paltan which he was going to deliver to his parents. Battalions still had proper systems in place to do justice to the deceased soldiers and enable his dependents to complete the last rite rituals as per their family customs.

It was again a long trek ahead and difficult to make it to POKHARA in one day. Hence, I had planned to halt at a village called MARJENKOT in KASKI District, just bordering LAMJUNG. This was the village from where two of our Battalion persons hailed and both happened to on annual leave. L/ Nk Jas Bahadur Gurung was in Signal Platoon and his younger brother Rifleman Santa Bahadur Gurung in 'C' Company. Their father was also a retired JCO from British era. I noticed a very interesting thing here. Retired Subedar Saheb used to conduct periodic classes on URDU language for

Muslim boys of the area. He had served in North-West Frontier Province for 28 long years prior to his retirement and acquired fair knowledge of Urdu for imparting to primary level students. It goes to prove that any type of knowledge gained is an asset and can be made use of on a future date, when required. So much to learn from the retired Subedar Saheb.

I had now left Lamjung District behind and travelling through Kaski District. Next day, I started for Pokhara bit late as it was just 4 hours march from village Marjenkot where I had stayed last night. I walked leisurely and took almost 6 hours. There were two beautiful lakes, named ROOPA TAL and BEGNAS TAL en route. By the time I reached Pokhara it was almost evening and stayed in a Hotel. Pension Paying Office at Pokhara used to be in thatched huts. Nepalese Govt had not permitted construction of any permanent Office or Residential building for PPO Pokhara since 1947 for some reasons. Whole complex gave a very shabby and uncomfortable look. But the Staff at PPO was always very cooperative and did their best for the Welfare of Gorkha Ex-Servicemen. Whenever I discussed this aspect with any retired soldiers in their villages deep in interiors or while walking in the country side, there was never any complaint from any one and PPO Pokhara was always held in high esteem by all Gorkha Ex-Servicemen.

# SNIPPETS FROM TENURE IN MIZORAM: 1975 - 76

## NIGHTMARE AT NGUR

Our Battalion was comfortably ensconced at a placed called HAYU-LIANG in LOHIT Sector of Eastern ARUNACHAL PRADESH since induction in June 1973. All of sudden, orders were received to move to Mizoram as there was a major spurt in insurgency, to the extent that Inspector General of Police and Superintendent of Police at Aizawl both were shot dead by the Mizo insurgents in their respective offices in the State Capital. We were inducted in Mizoram in February 1976. Battalion HQ and support elements were deployed at a place called Durtalang, just short of Aizawl and all four Rifle Companies sent to four different corners of Mizoram; a Company each at Darlawn in the North, Champhai in the East, Lunglei in the South and Kwartha Tha Weng in the West.

I was a Captain and sent with 'B' Company at Champhai in the East, adjoining Burma. I occupied a Platoon Post at a village called ZOTE which was 5 km North of Champhai. Sometime in Jul 1975, I took out a Patrol from Zote to Hnahlawn which was further, approx 10 km North on border with Manipur. Some troops of our Battalion and elements of ASSAM RIFLES were deployed at Hnahlawn. This area had been a major hotspot during previous round of insurgency which took place ten years ago in 1966-67. Troops of 9 BIHAR, then located at Hnahlawn Post, lost two Majors and numerous soldiers in a raid by the insurgents at that time.

Hnahlawn was approx a day's march from Zote and two villages named NGUR, pronounced as NOOR, and VAPAR were falling en route. Village NGUR was approx four km from Zote and at the highest point on a ridgeline overlooking the entire area. Village

VAPAR was comparatively smaller village and located 3 km further ahead.

Patrol reached NGUR village around 11 am. Since the village was at a dominating height, the movement of any Patrol could be seen from top of the village well in advance. I found GAON BURHA and few other villagers already gathered to welcome us. We had an ASSAM RIFLES soldier as an Interpreter with us who was proficient in Mizo language. We spent half an hour with them. Before departing from the place, I asked GAON BURHA if we could be of any assistance to the village. He hesitatingly requested me to give a SHAVING BLADE. Since I expected a request for rations or medicines to come from him, this took me by complete surprise and I, equally hesitatingly, asked why the request for SHAVING BLADE. The reply which he gave shook my very conscience and the same is etched in my mind forever. He said, "Sir, a young girl in the village is undergoing labour pains but unable to deliver the child as yet. In case the child is not taken out of her womb soon, child and mother both will die". He wanted to cut open the womb himself and thus at least save the child. I offered to evacuate the expectant mother to District Hospital at CHAMPHAI manually with the help of our soldiers. Since the Lady would not have survived the ordeal of long manual evacuation in steep hilly terrain, it was not accepted by the village elders. Patrol, thereafter, moved ahead for our next destination. On our return few days later, we again passed through NGUR village and I enquired about the fate of the Lady. I was told that it was too late to even cut open the womb and save the child at least. Neither the child nor mother could be saved.

Medical and other basic facilities in those remote and far-flung areas of our North-Eastern region were very primitive, almost non-existent during those days. District Hospital at Champhai was only for namesake and had only very basic medical facilities. People sitting in comforts of Metro Cities cannot visualise the difficulties faced by our hapless brethren in those un/under developed areas. It is this protracted neglect by the authorities which drives its inhabitants to pick up Arms against the very State they belong to.

## COMMUNITY SHARING OF SURPLUS FOOD ITEMS

During my next visit to the village NGUR, I noticed a small wooden Platform erected at the centre of the village. Size of the wooden Platform was approx 3 ft x 3ft and raised to approx 5ft height with 2 – 3 bamboo steps provided to reach it. I noticed some cooked food and other eatable items placed on the Platform in a haphazard manner.

I enquired from the GAON BURHA the very purpose of this arrangement. Politely, he explained,"Sir, all villagers put their surplus cooked food and other eatable items at this Platform every day. Any needy person is welcome to pick up items of his requirement from this place". What a great community sharing arrangement arrived at by these Tribal people! Surplus food items from each house get shared amongst the needy people all the time and without any hesitation. And, we call these Tribal people as BACKWARD; instead we the city dwellers are backward in this regard. May God grant SADBUDDHI to us on such like matters.

## NIGHT HALT AT VAPAR VILLAGE

VAPAR is a small village between NGUR and HNAHLAWN villages. Earlier there used to be an ASSAM RIFLES Post which had been vacated and the area lying deserted. During my Patrolling days at ZOTE Post, heavy rains forced me to take night halt at this village. I took shelter in the thatched hut of a lonely old man and remainder Patrol occupied the deserted Post of ASSAM RIFLES. There was a central fire place in the hut for cooking and other domestic purposes. It was raining down very heavily that night. Bay of Bengal storms during rainy reason are always very challenging. Pouring down a bucket full of water on one's head could be the nearest equivalent to clouds emptying water in the surrounding areas.

I had my emergency rations – SHAKARPARAS – with me and took them as my dinner. I noticed an iron net placed at approx two feet above the fireplace and certain things, not recognisible by me, placed over it. It was sufficiently higher so that fire does not reach the contents placed over the net directly yet the smoke emanating from the wood burning below could condition them appropriately for their usage at a later stage. Old man pulled out one such item placed above the iron net, cut it into many pieces, applied some salt and locally grown chilly powder. He had already boiled the rice and his meal was ready to eat.

Next day, before the commencement of our move ahead, I spoke to our Mizo language interpreter accompanying the Patrol as to what that contraption was which the old man had prepared for his dinner. He hesitatingly told me it was a fully skinned-out Field Rat whose skin part alone had been retained and smoked for longer life preservation. Smoked skin of the field rats was cooked by the locals when other food items were not easily available.

## FATHERLESS CHILDREN'S HOMES AND NOOLARIM

Mizoram has a typical social institution known as FATHERLESS CHILDREN'S HOMES. It is established at many places all over the State. I was quite surprised to notice a board hanging at the entrance of a small building in the Market place in Aizawl displaying existence of such an institution. When enquired as to what this Institution was and why its necessity had arisen, where from these fatherless children had arrived, we came to know of an equally interesting social custom prevailing in the Mizo Society, known as NOOLARIM which happened to be the root cause of necessity of opening such like Social Institutions.

What is NOOLARIM? It is a social custom prevalent in entire Mizo social structure where in majority of young boys and girls get together at a central place on every Saturday evening, enjoy sing-song, dancing, drinking whole of Saturday night and Sunday and return to their homes on Sunday evening or may be even Monday morning. It is a no-holds-barred social meeting of only young unmarried people and has full acceptance of their family elders. Hence, crossing the limits of a social get together into a rather opportunistic and promiscuous relationship was much likely to happen, more so when it has the encouragement from family elders. Strange as it may appear, it was quite normal for

Mizo Ex-Servicemen to visit our CSD Canteen at Durtalang and buy Cosmetics items for their daughters so that the girls prepare themselves well for the weekly NOOLARIM event.

With so much of freedom to young boys and girls, birth of children 'out of wedlock' was but natural. Also, no girl or her family members were ready to keep the children born out of NOOLARIM. There were numerous factors for non acceptance of the children; difficulty in ascertain the paternity of the child, unwillingness of the male partner were only few of the reasons. Thus arose the requirement of establishing FATHERLESS CHILDREN'S HOMES. This practice of NOOLARIM and prevalence of FATHERLESS CHILDREN'S HOMES was there in Mizoram when our Battalion was deployed in Mizoram during Counter-Insurgency Operations in 1975 – 76. Not sure of its prevalence now, almost half a century later.

Closest to this system of NOOLARIM of Mizoram, there is a practice of RODIGHAR in the villages of Western Nepal. In this too, young boys and girls of the village get together at a house on the outskirts of the village every Saturday evening, sing-song, drink, dance, thus enjoy the whole night and return to their homes on Sunday. In this case too, children are born out of this union of unmarried boys and girls. But Nepalese social structure has resolved it in a different manner where no requirement of opening any FATHERLESS CHILDREN'S HOMES arises. In case the girl gets pregnant, the child is kept by the girl and the boy with whom she marries subsequently, becomes the father of this child born out of RODIGHAR union too. Neither there is any hesitation on part of the girl to have the child before marriage nor in the mind of the boy with whom she gets married subsequently. Yes, the child is often known in the village as RODIGHAR WALA.

## A SIAMESE CAT

While I was Post Commander at ZOTE Post, one day my Post Havildar caught a Burmese PAITE and brought him before me. PAITE is the name of a Tribe which resides 20 km to 30 km both sides of international border between India, in this case Mizoram, and Burma. This Tribe has common social structure, customs and traditions on both sides of International Border; even marriage between Indian and Burmese PAITE families was common. Many of these people were regular conduits for illegal trade of smuggled goods and other contraband items. It was a routine practice to catch the PAITE crossing over to India and hand them over to the Civil Police.

However, the PAITE brought before me that day was a bit different. He did not carry any smuggled/contraband goods with him but, of all the things, he carried a Tiger Cub. The animal was approx one and a half foot in height and a beautiful creature to look at. I was fascinated seeing it and immediately decided to buy it. I paid him whatsoever he demanded and organized a largish wooden cage for it. I put the Cub under overall supervision of the Post Havildar who was instructed to buy one kg of raw meat everyday from ZOTE village and feed it.

The Cub, though not very friendly to any one, was a very active animal. Soon the word spread in the entire village and a number of villagers used to come to the Post just to have a look at it. Especially for village children, it became an amusement centre. Many persons from other nearby Posts also used to come and see it. Few times, local village dogs wandered near the Post and noticed the Cub. However, when the dogs came very near the Cage, one short hissing annoyance was sufficient for the dogs to scoot and never to return again. We all used to enjoy the spectacle of village dogs running away and gave me reason to feel proud that the Cub belonged to me. Slowly, my Commanding Officer and other officers at Battalion HQ also came to know of it. Almost four months had passed since Cub had been at Post and we noticed no perceptible change in his body growth. It looked exactly the same what it was when bought it from the PAITE. I even thought of taking advice from some Veterinary Doctor in Aizawl in this regard.

Meanwhile, there was change in Command of the Battalion. Lt Col Y C Bali went on posting and Lt Col Vinod Badhwar assumed the command of the Battalion. New CO soon visited our Post at Zote and I, very proudly, showed him my most prized possession, the Tiger Cub. He had already been briefed about it in the Battalion HQ and was well aware. After having an intense look at the animal, he affectionately put his hand on my shoulders and said," You Bloody R V Singh, it is not a Tiger Cub, it is a SIAMESE CAT. I suggest you set it free in the jungles earliest".

I was left devastated. After all, I had spent money on its purchase and feeding it for last four months. And now, I have to let it go in the jungles! All my dreams of owning a Tiger Cub came crashing down. Good old dictum of "Man proposes and God disposes" is so very true.

# EXEMPLARY GRIT AND DETERMINATION

It happened sometime in May/June 1980. 1/5 Gorkha Rifles was located at a place called Pallanwala, near Akhnoor in J&K. Lt Col (later Maj Gen) I A J Cardozo was commanding the Battalion and I was doing his Adjutant for a while. A short introduction of this great personality is needed before we proceed further. General Cardozo was commissioned in 1/5 GR in late 1950s, served with the Battalion for 5 years and then posted to 4/5 GR on its re-raising in 1963. He led a Rifle Company in the 1965 Indo-Pak War in J&K and was again was leading a Rifle Company in 1971 Indo-Pak War, this time in Bangladesh, where unfortunately, he lost his leg in a minefield. He cut his injured leg with a KHUKRI himself before the wound could cause any life endangering damage. He, a War Casualty, fought a long drawn battle, this time with his own organization, that is, Army HQ, for getting command of an Infantry Battalion. He was given Command of 1/5 GR, thus setting a precedence for many others to benefit in future.

I saw him as my Commanding Officer with the eyes of an Adjutant. A thoroughbred professional and very straight forward person. Despite his leg amputation, he attended all training and administrative activities, participated in Sports, Swimming and regularly visited all Posts occupied by the Battalion. Physical disability could not deter him even a bit; rather it appeared to be motivating him to achieve the impossible; which he did. There was so much to learn from him in those crucial formative years of mine. I wish I had a longer tenure with him.

It happened sometime in June-July 1980. We were conducting a 2 Mile Run as part of periodic BPET for the entire Battalion. Commanding Officer announced to not only participate in the Run but run the entire length of 2 mile with Jawans. We were worried about his physical well being as running for a distance of 2 miles will definitely injure his amputated leg thus worsening his health. He was adamant to run and all our efforts to convince him otherwise bore no fruits. We were getting worried. For some divine intervention, he changed his mind and agreed to run only half the distance but run he will. We had some solace.

Run commenced next day with Adjutant standing at the Start Point and Commanding Officer at the Turning Point which was exactly 1 Mile away from the Finish Point. He ran for the entire distance of 1 Mile with an amputated leg and completed it in good time. How difficult it must have been for him throughout the Run!! His amputated leg had been badly bruised and bleeding and took a number of days to recover. What a grit and determination he had to go through all the travails and tribulations of life since the day he lost his leg in the minefield during Indo-Pak War in1971! A truly commendable feat.

I have been blessed to serve under many such great Commanders in first ten formative years of my service. To quote them and in time chronology; Maj Gen S C Gupta, Brig Y C Bali, Maj Gen Vinod Badhwar, Maj Gen I A J Cardozo and Lt Gen V K Sood from DOGRA Regt who was Commander 33 Mtn Bde when I was posted on a Grade 3 Staff in the Bde HQ. There was so much to learn from each of them. Unfortunately, only two  of them ie Maj Gen Vinod Badhwar and Maj Gen I A J Cardozo are alive today and I pray to God Almighty for their good health and long life. May God grant SADGATI to the remainder three great souls who are no longer in this world.

# REPORT CARD: D S S C-37 (YEAR-1981)

I was rather fortunate to attend Course Serial – 37 at Defense Services Staff College, Wellington (Tamil Nadu) in the year 1981. Very professional environment and equally matching administrative support to all attendees. Perfect opportunity to make numerous life-long friends. Lot of MARA-MARI for the serious souls but a heavenly one year long sabbatical for those not so serious ones and a few newly married couples too. We had few Emergency Commissioned Officers attending the course who were almost 8 - 9 years senior to us and even stood approved for the next higher rank. We had an officer from Indian Air Force who was already a Wing Commander while attending the course, retained as Instructor at DSSC on the conclusion of the course and subsequently rose to become Chief of the Air Staff. We five officers from our course were rather young, though one amongst us had figured in Competitive List also. I doubt if anyone seriously noticed our presence on the course. As per DSSC norm, on conclusion of the course, we all were interviewed by the Commandant. My interview lasted for barely a minute in which he congratulated me on my posting as Instructor Class 'B' as I was going on promotion from Captain to Major where as he was not so lucky 30 years ago and sent on posting as a GSO3 (Operations ) in a Division HQ, an appointment tenable by a Captain only.

We all attendees of DSSC – 37 have retired from Service long ago. It will be interesting to go over how the destiny treated all of us. Here it goes, with malice towards none, dead or alive.

We were a total of 285 students who attended the course. One Army Officer (Captain S R Savant, 5 GR/JAKLI ) was returned to his Battalion on medical grounds midway through the course. Out of these 285 numbers, Army had 175, Navy – 35, Air Force -50 and 30 students from friendly foreign countries. 20 Army Officers from the 175 mentioned above, were placed in Competitive List, out of which 5 Officers went to attend Courses at Foreign Staff Colleges. Details of Officers mentioned below includes these 5 Officers also.

Now, let us see the respective levels all student officers reached prior to their superannuation from Service.

1. **LT GENERALS** – 20
   S P Raman, ENGINEERS
   Daljeet Singh, ARMOURED
   J K Mohanty, DOGRA
   H S Panag, SIKH/ 5 GR/MECH INF
   H S Lidder, PARA
   P K Singh, Artillery
   P K Gangadharan, MARATHA LI
   B J Gupta, 11 GR
   J S Lidder, GRENADIERS
   A S Bajwa, ARTILLERY
   B S Yadav, BIHAR
   C B Vijan, ARTILLERY
   A S Sekhon, SIKH LI
   D D S Sandhu, AOC
   K K Kohli, ARTILLERY
   S P Sreekumar, SIGNALS
   Avadhesh Prakash, NAGA
   **Attended Foreign Staff College Course**
   Sudhir Sharma, GUARDS
   R K Chhabra, 4 GR
   P K Samantray, ARTILLERY
2. **MAJOR GENERALS** - 23
   M P Bhagat, MAHAR
   Abhradhwaj Parmar, RAJPUT
   Arun Roye, GUARDS
   U K Bapat, KUMAON
   Harwant Krishan, 4 GR/MECH INF
   R T Thadani, ENGINEERS
   B S Himalayan, ENGINEERS
   R S Jambussarwala, ARTILLERY

S K Jaswal, ENGINEERS
KS Randhir Sinh, 3 GR
R K Singh, RAJRIF
T K Kaul, KUMAON
Lakhwinder Singh, ARTILLERY
R K S Shivrain, PARA
R P S Malhan, PARA
M S Moorjani, AOC
Dipak Mukherjee, BIHAR
N S Pathania, 5 GR
P P Rajagopal, PARA
P S Rana, GRENADIERS
S K Sahijpal, ENGINEERS
M P Singh, SIGNALS
Rajpal, ASC

3. **BRIGADIERS** – 47
4. **MAJORS TO COLONELS** - 85
5. Died of Cancer - Major Gulshan Singh, KUMAON
6. Died in Heptr crash (with Lt Gen Jameel Mehmood ) – Col Darshan Singh, SIKH
7. Died of Alcoholism – Lt Col A A Khan, ARTILLERY
8. Court Martial / Serious Cases Investigation
   Lt Gen – 1 ( Avadhesh Prakash – case later set aside )
   Maj Gen – 1 (T K Kaul )
   Colonel – 1 (D D Pawar, ASC; figured in Competitive Vacancy list )

It had been an interesting one year of life spent in the lovely environs of Nilgiri Hills. I had the opportunity to interact with great faculty members and equally great colleagues with whom God gave me an opportunity to attend the course. May God bless all of them.

# INTERACTION WITH FIELD MARSHAL S H F J MANEKSHAW

Field Marshal S H F J Manekshaw was a great soldier and a great human being. The impression he left on the mind of a 2/Lt ( myself ), when he visited Khemkaran Sector in November 1971, just a week prior to Indo-Pak War, is indelible even today. I saw him climbing the stairs of a Lecture Stand on Ditch-Cum-Bund, faster than the Army Commander, Corps Commander and Divisional Commander. He looked physically very fit, smart, perfectly attired, exhibited positivity all around and motivated everybody who came in contact with him that day. I, a very junior officer, a 2/ Lt, was completely overawed by his personality.

I was fortunate to have another good interaction with him a decade later, this time while doing Staff Course ( D S S C - 37 ) at Defence Services Staff College, Wellington in 1981. After his retirement, Field Marshal S H F J Manekshaw had permanently settled down at Conoor, a small Hill Station just adjacent to D S S C, Wellington, near Ooty. It was a custom followed by all officers of Gorkha Regiments attending each Staff Course, as a group to invite Field Marshal for a Social function at local Gymkhana Club and, in reciprocation, he will invite all Gorkha Officers, as a group, for drinks at his Residence.

He was duly invited and arrived at Gymkhana Club dot at the given time. As expected, he was the most sought after person all through the Party, everyone vying to get photographed with him and stand close to him for maximum duration. His conversational skills, delivery of anecdotes from his military life and cracking of spontaneous jokes was just too good. Around 11 pm, Field Marshal desired to return to his home. We all were there at the Departure Lounge of the Club to see him off. Suddenly, Field Marshal noticed a beautiful and extremely charming lady whom, somehow, he had not come across during the whole Party. He reached out to her, put his hand on her shoulder and we all returned to the Main Hall of Gymkhana Club. Party got extended by another hour. We all enjoyed the extended part also of the Party.

Lady was wife of a Captain attending the course. No doubt she was a beautiful, educated, intelligent and very well behaved Lady. Captain was also a smart, intelligent and indeed an outstanding Officer. Even while undergoing the Staff Course, his friends and many others attending the course used to say that he will one day rise to become an Army Commander, if not Chief. Yes, the Captain was bit egoist and kept aloof from even his Course mates for the entire course duration.

Captain surely, as expected, rose to become an Army Commander. The Lady had the honour to come to D S S C Wellington again, this time with her second husband who attended Staff Course few years later and retired as a Colonel. Such is the journey called 'life'. May God bless all of them always. We extend our most respectful and sincere good wishes to them.

# CURVES OF CAREER - MAJOR HIRA LAL THAPA

Hira and me had known each other since our students days in Dehra Dun. He was doing his Post Graduation in Physics and myself in History before joining Army. He was basically a fine person and belonged to Garhi Cantt in Dehra Dun. We both were selected for Commission in Army together and joined Officers Training Academy, Madras in November 1969. In Course strength of 530, Uttar Pradesh had contributed 80 and Dehra Dun alone could boast of 22 out of this 80. Most of us knew each other very well. Training schedule at Academy was so hectic that we all almost lost track of each other and did not even realise when First Term was over.

Second Term was in progress. Boxing was an important Sport and matches at Inter-Company level were going on. Hira in any case was a good Boxer and we all from Dehra Dun had high expectations from him. Destiny was not in his favour and, in one of the final matches, he suffered a serious blow to his head. Hira fainted in the Boxing Ring itself and had to be soon evacuated to Military Hospital. He remained unconscious for days together as head injury had caused serious Neurological disorder in his brain. Process of recovery was rather slow and long-drawn. We all completed our training and passed out from Academy; dear Hira was still in Hospital with a ring of uncertainty hovering over his head. We all wished him a speedy recovery and left Madras for our respective Battalions. Hira remained in Hospital for another one year before he was declared fit to undergo remainder part of his training. He lost 2 Terms and with his Physics background, was commissioned in the Corps of Signals.

We did not come across each other for next 13 long years and, truly speaking, I had lost track of him. In early 1985, I was serving with my Battalion at Roorkee and was detailed as Member of Board of Officers for conduct of Promotion Examination Part 'C' for Officers. Examination was to be conducted at Dehra Dun. A Brigade Commander from Dehra Dun had been detailed as the Presiding Officer. Myself and another Major who happened to be good friend were two Members of the Board. More than 50 Officers from various Units and other Establishments all around had gathered to appear in the examination. We two Members were having a cursory look at the gathering and I located my good old friend Hira standing in a small group of Officers. I was thrilled to see him after such a long time, could not restrain myself and straight walked over to him. Hira also felt very happy seeing me after such a long time. After exchange of normal pleasantries, Hira asked me, "I thought I am late in clearing Part 'C' Exam, had two unsuccessful attempts, how come you have not done it so far?" I realized Dear Hira had mistaken me to be one of the candidates for appearing in the exam. I did not want him to feel embarrassed and told him it was owing to certain unavoidable circumstances and hope to pass this time. Also, I understood the gravity of situation in his case. As he has not been able to clear Part 'C' exam till now, it would be difficult for him to do it in future too. I should somehow help him out. Also, I knew it was not a competitive exam and his passing will in no way affect the career prospects of other Officers.

Presiding Officer was known to be a stickler of the laid down Rules and Regulations. I had seen him as CO of an Infantry Battalion for two years while I was posted on a Grade 3 Staff in the same Brigade. I spoke to the other Member of the Board of Officers and explained him all aspects of the case. Also, I told him that unless we two Members give him 100% marks, my friend Hira will surely once again fail as he had done in his previous attempts.

Let us now cursorily go through the methodology of conduct of Part 'C' Exam. Entire exam has three Parts of a total of 200 marks. Part 1 consists of a short tactical appreciation and is conducted by Presiding Officer and allocated 120 marks. Part 2 consisted of few questions on administration and had been allocated 50 Marks and conducted by Member No 1, self in this case. Part 3 consisted of some questions on Map Reading and had been allocated 30 marks and conducted by Member No 2. Passing marks were 40% ie 80 out of total allocated 200 marks. I was very definite that my friend Hira had been getting low marks in Appreciation, hence, unsuccessful in all previous attempts. I explained to Member No 2 that unless we both give 100 % marks left at our disposal to Hira, he is bound to fail this time too. He very kindly agreed to help Hira.

Examination was over by the evening. Presiding Officer handed over his marks sheet to me and asked to compile the results and show him again and before declaring the results. I checked up the Marks scored by Hira in Appreciation part of the exam. As anticipated, dear Hira had scored NIL marks in Appreciation part of the Exam conducted by the Presiding Officer. Since we two Members gave Hira full marks, he scored a total of 80 marks, minimum essential to pass in the exam. We declared the results in which my friend Hira also qualified. I never told him anything. I went over to Dear Hira again, congratulated him and we both departed - never to meet again. He had been posted to a Composite Signal Company around Ranikhet and joined his Unit.

I suppose Hira could not succeed in clearing his Part 'D' Promotion Exam. He retired early and settled down at Garhi Cantt, Dehra Dun. Some time ago, I visited Dakra Bazar in Garhi Cantt, his original place of residence in Dehra Dun and made an attempt to locate him. It was sad to learn that my friend HIRA LAL THAPA had already expired few years ago. We were destined never to meet again. May God grant SADGATI to the departed Soul.

# A BRILLIANT CAREER WENT STRAY - CAPTAIN JEEVAN JOSHI

Jeevan Joshi and me were friends from our College days in Dehra Dun days since mid 1960s. Short in structure and thinly built, though a KUMAONI Brahmin but looked more like a Gorkha from Nepal. We joined Officers Training Academy together in the end 1969 and were commissioned in the same Regiment, 5 GR, I joining the First Battalion and he Sixth Battalion. Soon after commissioning, we attended Young Officers Course together which was more like one more Term of pre-commission training flung on our shoulders; same bonhomie and could not care less atmosphere continued for five more months. Then came 1971 Indo-Pak War which we all fought as a Rifle Platoon Commanders. This gave us lot of confidence in ourselves and much to brag about our experience in War. In July 1973, we all were again back at Infantry School, Mhow, this time to attend Battalion Support Weapons' Course of five months duration. Jeevan and me decided to share a room together for the duration of the course. Thus our friendship further grew and we understood each other better. He was intelligent and hard working but rather emotional, perhaps little more than required.

Jeevan married a Gorkha girl from Garhi Cantt in Dehra Dun in mid-1972 when he was a 2/Lt and had left her at her parental for the duration of the course. However, he used to miss his wife all the time. He got into the habit of drinking every evening and I noticed him on the Bar in Student Officers' Mess almost daily. Often I had to pull him out from the Bar to have his meal. Saturdays and Sundays used be more troublesome. It was becoming difficult for him to beat separation from his newly wedded wife.

We were undergoing RCL Leg, the final Leg of our Course and just a fortnight away from the conclusion of the course. It was a Saturday evening. I found my friend Jeevan sitting on Student Officers' Mess Bar and gulping down drinks one after another. He was almost drunk. When I asked him to move for Dinner, he started crying. He had received a letter from his wife informing him of her being seriously ill and admitted in Hosp. My friend Jeevan had already reached his breakpoint and could no longer bear separation from his dear wife. Next day was Sunday, we discussed the matter and he decided to be with his wife during her sickness at all costs. He met Senior Instructor of RCL Wing, named Major Nathu Singh from PUNJAB Regiment and got four days short Casual Leave sanctioned. He was dropped at Indore Airport the same day by a colleague. That was last I saw of my friend Jeevan Joshi and could not even imagine I will never see him again. He was to report back by Friday morning. Our course finished 9 days later; he never reported back and became a DESERTER from Army. I collected all his personal belonging, deposited with Infantry School, Mhow and returned to my Battalion at Hayuliang in Lohit Sector of Arunachal Pradesh.

I never forgot Jeevan, memories of his association with me always lingered in my mind. How could Jeevan desert from Army! I just could not comprehend. Separation from ones dear wife does not mean one should desert from your place of employment. It was later learnt that he and his wife took an emotional decision for Jeevan to leave Army forthwith even if it amounts to desertion. Both fled to Kathmandu in Nepal where some relation of his wife supposedly held a position of authority.

I was proceeding on three month long trek to Regimental Recruiting Areas in Western Nepal in October - December 1974 and decided to meet them at Kathmandu. First, I visited his in-laws place, named KARKI KUTI in Garhi Cantt, Dehra Dun, met his father-in-law and took their address at Kathmandu. Those days Jeevan had been employed with YETI TRAVELS at Kathmandu. I tried to trace him but could not as he had shifted to some other unknown place. It was also learnt that Jeevan had been contacted by Chinese agents too. He was subsequently arrested by Nepalese Police and handed over to Indian authorities. Later, sometime in 1975-76, he was brought to Mhow, duly court-martialled and 'Cashiered Out from Service'. He went back to Kathmandu. I once more met his father-in-law at their Garhi Cantt House sometime mid-1980s. By then, my friend Jeevan had become a father of 3 daughters and was living in Kathmandu with his family.

Few years ago, I learnt of Jeevan having finally returned from Kathmandu to Dehra Dun. Alas! He did not survive long and died of some serious health complications sometime in 2016-17. I thought of meeting his wife, named Mithlesh and pay my condolences to her. Her parental place KARKI KUTI had long been demolished and replaced by new construction. I was able to meet his wife and her two brothers. I met the lady first time in my life and told her that her husband was way ahead of me in every respect of professional life as young officers and had he stayed in Army, he would have definitely risen higher than me. Her three daughters had been married and leading respectable lives. Her son had been recruited in 11 GR. His father's desertion from service and subsequent Court-Martial almost derailed him from the Training Centre. Army is an excellent organization and permitted the boy to continue in his endeavour to make a Career in Army according to his destiny.

I still consider Jeevan Joshi was much better than me in all respects. He was brilliant and almost headed for the award of best YO of our Course. For some strange reasons, no Officer from our course was conferred the award of Best YO; hence, Jeevan missed on that. He was doing exceedingly well on BSW Course too. Destiny had a different course lined up for him. His emotional imbalance turned the tide against him, and he died unsung. May God grant SADGATI to the departed Soul of my dear friend Jeevan Joshi.

# IRRESISTIBLE DIL OF DIL BAHADUR GURUNG

Rifleman Dil Bahadur Gurung joined our Battalion sometime in 1969 -70 in J&K and was assigned to 'C' Company. He was tall, smart and ever-smiling. His looks gave no indication that Dil Bahadur possesses an uncontrollable DIL(heart). This weakness was to land him in serious troubles a number of times in future. One could not fathom that this young Soldier, just 19 – 20 years of age and yet to be married, has an uncontrollable urge for the company of females. Battalion had moved down from J & K to Ferozepore in December 1970 and had soon settled down in routine tasks of a Peace Station life. In mid 1971, while part of Night Guard at Battalion Officers' Mess, Rifleman Dil Bahadur Gurung managed to enter in the house of a local Washerman and was caught trying to force himself upon the wife of the Washerman. He was promptly awarded a punishment of 28 days Rigorous Imprisonment. That was his First Red Ink Entry in his Service Records. Surprisingly, he repeated the same offence at the same place, same female next year and was soon duly punished with a Second Red Ink Entry punishment of 28 days RI. Immediately thereafter, he was sent on annual leave to Nepal to get married which he did.

Meanwhile, in mid 1973, Battalion moved from Ferozepore to a place called Hayuliang in Lohit Sector of Arunachal Pradesh. After few months of return from annual leave, Dil Bahadur was again at it. This time, he was caught being infected with STD, and as per existing Rules, he was awarded another punishment of 28 days RI, a Third Red Ink Entry. We asked him why did he go to the infected female. His reply may appear hilarious to many but was an honest burst of a truthful confession from an otherwise simple soul. He said, "Sir, I drank two bottles of local RAKSHI (brew) from her and paid the money. She did not have the change to return the balance amount and instead offered her services to me. I had no other option to get value for my balance amount". It took him six months of treatment before being declared fit for duties. But Dil Bahadur had an irresistable DIL and could always accommodate more than one at any time, any place. In end-1974, he was again caught by the parents of a local Mishmi girl, this time in their house in the village. Thus, he earned his Fourth Red Ink Entry in five years of Service in the Army.

As per Instructions in vogue that time, any soldier who has been awarded more than 3 Red Ink Entries, may be discharged from service irrespective of the fact whether he has completed pensionable service or not. However, before his case for discharge from service could be processed, in February 1975, Battalion moved from Arunachal Pradesh to Mizoram in an emergent manner and were so occupied in combating the Insurgency for the next one and a half years that his case for discharge from service got almost buried in the files and was lost sight, being of low priority in an intense insurgency environment. Also, he did not indulge in any such activity during the next two years. Thereafter, in mid-1976, Battalion moved from Mizoram to Trivandrum and by then hardly anyone even remembered the past exploits of Rifleman Dil Bahadur Gurung. Meanwhile, there was a change in command of the Battalion, Lt Col Y C Bali had moved out on posting and new CO, Lt Col Vinod Badhwar took over while we were still in Mizoram.

I was officiating as Adjutant in mid-1977. At mid night, I got a call from local Police Station to collect one of our Soldier who had been caught barging into the house of a civilian in the area with an intention to outrage the modesty of the Lady of the house. The person involved was none other than irresistable Dil Bahdur Gurung. This earned him a Fifth Red Ink Entry in 9 years of service and he was still six long years away from earning his Pension. Few days later, I put up his case for his removal from service to the CO which as per Rules was without any pensionery benefits. CO pended his decision and, in the meantime, scrutinised his service documents, spoke to Sub Maj and his Company Senior JCO. He took into consideration Dil Bahadur Gurung's young age, a dependent family consisting of his wife, two small kids and old parents. CO gave him another chance to remould himself so as to enable him to earn his Pension.

It was in the end 1983, full six years had lapsed since the above narrated incidence had occurred at Trivandrum. I had been posted as Instructor at Infantry School, Belgaum. Dil Bahadur Gurung was then posted in Demonstration Company with Infantry School, Belgaum. One day he visited my house. He had completed 15 years of service and was proceeding on Pension few days later. My thoughts went back to all occurrences of his past service in the Battalion, more so the humane and very mature decision of Lt Col Vinod Badhwar, then Commanding Officer, over-riding all existing Rules/Regulations on the subject which enabled Dil Bahadur to remould himself and earn his Pension. May God bless Rifleman Dil Bahadur Gurung.

# MY EXPERIENCES OF 1984 ANTI-SIKH RIOTS AT DELHI

## ONE

Mrs Indira Gandhi was shot at mid-day on 31 October 1984. Our battalion 1/5 Gorkha Rifles was part of Brigade at Roorkee and ordered to move to Delhi on the evening of 31 October 1984. We reached Delhi on the morning of 01 November 1984 and given a place to camp in the Parade Ground at Delhi Cantonment.

It was a pathetic sight to see the national Capital burning. We found there was at least one vehicle burning in a span of 300 meters to 400 meters on the road in Delhi. En route we saw numerous houses on either side of the road set on fire by the rioters.

We remained unutilised at Parade Ground, Delhi Cantonment for three full days and no orders for deployment to any place were received. Our Commanding Officer spent many hours every day sitting in HQ Delhi Area and waiting to receive executive orders for our deployment. However, neither Central nor Delhi Govt gave any orders for our deployment. Resultantly, riots continued unabated, and our entire Brigade remained unutilised. It was during these three crucial days of 01 to 03 November that major losses of lives and property of Sikhs in Delhi took place. Had the orders for the deployment of our Brigade been issued immediately on arrival in Delhi on the morning of 01 November itself, thousands of precious and innocent lives could have been saved and loss to the property minimised, if not completely avoided. Alas! It was not so done by the powers at the helm of affairs in Central Government and Delhi administration.

On the morning of 04 November, we were given orders for deployment in a specified area of South Delhi. Battalion HQ was set up in the Campus of IIT, Delhi and my Company in the area of Hauz

Khas with Company HQ located alongside the office of DCP, South Delhi.

## TWO

On 06 November, it was learnt that some major problem has erupted in the area of Gurudwara Nizamuddin. Local villagers had been indulging in arson and rioting against the Sikh residents of this area since the previous night. I moved there with a Platoon of Gorkha soldiers. A large contingent of Policemen was already at the scene. I noticed one Mr Kakkar, whom I later on learnt to be the Commissioner, Delhi Police, was also there and trying to control the situation himself. However, no bullets were fired by the Police to disperse or control the arsonists. Most of the Sikh residents of the area had taken shelter inside the Gurudwara. Rioters were trying to enter the Gurudwara complex and Delhi Police trying to prevent them. Seeing an Army contingent moving in, rioters slowly disengaged from the Gurudwara but were moving around menacingly.

Meanwhile, an old Sikh couple walked up to me from the crowd gathered inside the Gurudwara. Both of them were crying and requested me to rescue their daughter-in-law with their three years old grandson left behind in their house not far from the Gurudwara. They feared for their lives any time. I took three armed soldiers with me from my Platoon, which by then had been deployed around the Gurudwara and asked the old man to accompany me to their house. En route, we noticed rioters moving in the lanes so as to find out if some Sikhs were still left around. We reached the house. On seeing Army Soldiers with her father-in-law, the lady opened the door. I could well see the fear of death lurking in her eyes; she was so terrified for the safety of her young child and her own self. Even now, when I get reminded of her eyes ridden with fear of sure death, it gives me shivers. I felt so relieved and happy escorting her safely to Gurudwara. May god bless her and her young child, wherever they are.

## THREE

It was sometime in second week of November. Entire Delhi had been under night curfew from 8 pm to 6am. As expected, residents of Delhi did not understand the implications of a night curfew; for most of them, it was a great fun moving around under night curfew restrictions. They still behave in the similar manner even now. My Company was entrusted to enforce night curfew on the road leading from Hauz Khas to AIIMS. However, we noticed everyone moving as usual even after 8 pm Curfew time and none appeared to abide by the restrictions imposed on movement. Around 9 pm, my patience gave away and I put two Gorkha Soldiers in the middle of the road and told them to aim at the next oncoming vehicle and fire at the driver if he does not stop his vehicle or tries to speed away. It worked and in no time we had 10 – 12 vehicles parked on left hand side of the road.

Meanwhile, a Govt vehicle with national flag flying on its bonnet, screeched to halt. Apparently, there was some important Govt functionary inside the vehicle. I was standing close by and could well listen to the driver of the car trying to argue with the Gorkha Sentry who had stopped the car and the person sitting inside telling the driver not to argue with the Sentry. Since it was dark I could not recognise who was inside the car and was happily enjoying the incident. Later, I walked to the car and had a look to see who was inside the car. It was Mr Balram Jakhar, then Speaker of Lok Sabha, sitting alone behind. I respectfully told his vehicle had violated the curfew regulations as laid down by his own Govt. He profusely apologised. I permitted the vehicle to move on with a stern warning to the driver never to argue with an Gorkha Sentry lest he wants to get shot.

It was after 20 years in 2004, that I again happened to meet Mr Balram Jakhar. I was functioning as Director, Ex-Servicemen Welfare, Madhya Pradesh Govt at Bhopal. Mr Jakhar was then Governor of Madhya Pradesh. During the annual Flag Day function organised at Raj Bhawan, I reminded him of this incident. He well remembered it, had a hearty laugh and affectionately thanked me.

Mr Balram Jakhar, a senior politician during Mrs Indira Gandhi regime, was an astute politician and yet a gentleman. He died few years ago. May God grant SADGATI to the departed soul.

**FOUR**

The severity of the riots and arson was much more during the first two weeks of November and, thereafter, situation was brought under effective control. The incident I am now narrating occurred in 2nd week of November. It was learnt that some major tragedy had occurred at Tughlakabad Railway Station. A train coming from South to New Delhi had been force-stopped at Tughlakabad Railway Statiion by the rioters and few Sikh passengers killed. I took a Platoon strength of soldiers and rushed to the site.

Alas! By the time my Platoon arrived at the Railway Station, damage to the lives had already been done. I found few dead bodies still burning at the Platform itself. A huge gathering of rioters, could be in many hundreds, was still at the far end of the Platform and fast retreated on seeing arrival of the Army Soldiers at the scene.

I was shocked to see two half burnt bodies, both in Olive Green Army Uniform, lying on the Platform. I could identify them to be as officers from their half-burnt cloth epaulettes worn on their shoulders; one of them was a Lt Col and other a Captain. They were travelling in the train in Army Uniform from their Units somewhere around Jhansi and coming to New Delhi, forcibly taken out from the train by the rioters and killed. I could still see the Army boots, though in half-burnt state, worn by them. It was so pathetic, so terrible a sight which shook my very conscience to the hilt. I felt so apologetic for not been able to save the lives of even my own Army brethren. Two precious lives snuffed out by mob frenzy for no damn reason. Most unpardonable an act. May God grant SADGATI to the two Unknown Warriors.

**FIVE**

During 1st week November, I saw a middle-aged Sikh gentleman often visiting our Battalion HQ in IIT, Delhi and discussing something with the Adjutant. Being rather busy, Battalion could not attend to him on priority. In 2nd week, I again saw him in our Battalion HQ and asked him what his problem was. He was employed in the Security Deptt of IIT, Delhi and living with his family in IIT Campus. On 31 October 1984, his son, along with his friends, had gone to see Mrs Indira Gandhi at AIIMS since she had been admitted there on being shot by the assailants. An adolescent curiosity, I suppose. However, the crowd at AIIMS and other places in Delhi, got violent resulting into large scale riots all over by the afternoon of that day. Somehow, it became difficult for this young Sikh boy to return safely from AIIMS to his home in IIT, Delhi. His close friend took him to his house in Trilokpuri in East Delhi where his parents, a GUPTA FAMILY, lived. Trilokpuri till then was not that violent and became worst affected in entire Delhi a day later.

Parents of this Sikh boy received telephone calls from Mr Gupta informing him their son was safe. He had been requesting to arrange for early evacuation of the child as rioters had been regularly attacking his house to harm the child. So far he had been able to thwart all attempts of the rioters and may not be able to withstand for an indefinite period. Since East Delhi was not in our area of responsibility, his request remained on little low priority.

I spoke to my CO and sought his permission to rescue the child. He immediately gave the permission and asked me to exercise due caution. I took 6 – 7 Gorkha Soldiers on two vehicles with a Light Machine Gun Mounted on the front vehicle and soon moved out. I took the father of the child along with me for easy identification of the address. It was a terrible site en route, a lot of vehicles and houses burnt down all along and few dead bodies could still be seen lying on the road in the Trilokpuri area. I could see few APCs of some Mechanised Infantry Battalion deployed in the area.

It took us about two hours to reach the house of Mr Gupta where the boy had been given safe shelter for over a week under difficult and trying conditions. It was learnt that the rioters had made three attempts to pull out the boy from his house. However, Mr Gupta stood firm and told them very resolutely that rioters have to first kill his entire family before harming this Sikh boy. We thanked the BRAVE MR GUPTA and brought the boy safely to his house in IIT, Delhi. It was so delighting to personally experience that HUMANITY still exists and even under most trying conditions. Mr Gupta refused to hand over the Sikh boy to the rioters, thus even risking the lives of his entire family in doing so. May God bless the young Sikh boy, his parents and entire Gupta family.

Yes, the Battalion was duly rewarded for its performance. Central Govt declined to give any award as the awards to the conferred to the all ranks of Armed Forces during OPERATION BLUE STAR had not gone down well with the Sikh community. However, COAS decided to acknowledge the efforts of troops employed during these Riots, by powers as conferred on him by the Central Govt. In due course, the COAS awarded his Commendation to various individuals. Five COAS Commendations were awarded to our Battalion, out of which three awards came to my Company. It was a very nice feeling as efforts put in during those two months of Nov and Dec 84 had been acknowledged.

# RESOLUTION OF A SUICIDE INCIDENT

An unfortunate yet interesting incident happened when I was DQ of an Independent Infantry Brigade Group at Babina Cantt during 1987- 89. It was unfortunate as it cost life and interesting as one had to resort to some unorthodox thinking, devise unconventional means and take an otherwise considered 'avoidable risk' to one's career to arrive at an uncommon resolution to benefit the dependents of the deceased soldier. Successful resolution of this incident encouraged me to always have full faith in the Organisation as well as in one's destiny.

A Havildar from Brigade Signal Company, who hailed from a village in Meerut District, committed suicide one night by jumping in front of the running train in Babina Cantonment. Those familiar with Babina Cantonment will recollect that Railway Line divides Babina Cantonment into two halves; Eastern half called Old Cantonment where Infantry Brigade Group was located and Western half called New Cantonment where all three Armoured Brigades were located.

A new Officer had recently taken over the Brigade Signal Company and was still in the process of fully settling down. Occurrence of this incident had obviously rattled him. It was revealed from the colleagues of the deceased person in Signal Company, he had a serious marital discord with his wife. In that, he suspected his wife of having illicit relations with some person in his village. The deceased person had two growing up dependent children and aged parents to look after back home. He had gone on annual/ casual leave to his village three times during the preceding six months to resolve the matter with his wife and was under a lot of mental tension. This ultimately resulted in his committing suicide.

Those days Family Pension was not authorised to the dependents of the persons whose death occurred due to committing suicide. This was irrespective of the fact whether they have completed minimum service as laid down for entitlement of Pension or not. Hence, his dependents were not entitled for any Family Pension. This was quite a heavy price for his children and aged parents to pay. Hence, before initiating INITIAL REPORT on this occurrence of this incident to the Higher H Q / Army HQ, I thought over if we could somehow help his dependents in grant of Family Pension.

Brigadier S S Kaler, MECH INF /JAT, was the Brigade Commander. He was a very positive and readily decisive Commander. I discussed the entire case with him and my recommendation to help out the dependents. We planned to show the deceased person participating in a periodic Brigade Signal Company level Telephone Cable Laying Training Exercise to be conducted by Night. Necessary post-dated documentation, in this case, had to be done at Brigade HQ and Brigade Signal Company level. Brigade Commander fully understood the requirement and instantly gave the necessary permission to proceed ahead. Accordingly, documentation incorporating the above details was done and INITIAL REPORT and DETAILED REPORT forwarded to all concerned. However, I kept my fingers crossed lest it gets found out by someone in the chain. In that case, I will be the first person to face the serious consequences.

Lo and behold! After 5 - 6 days, I got a call from HQ Central Command, Lucknow. Brig D G S Gill, Artillery, Brig 'A' at HQ Central Command, had smelt the rat in this case and was on the other end to enquire further details from me. He straight came on the subject and asked me, "Son, tell me what this entire suicide case is all about". I told him every detail of the case, including our intention of helping the dependent children and aged parents of the deceased person, prior permission of the Brigade Commander and no foul play involved in the death of the deceased person. A great person as Brigadier D G S Gill was, he fully held our hands in the case and even spoke to our Brigade Commander appreciating my action.

In hind sight, what do I think of this incident; primarily of laid down Rules / Regulations and the roles played by Senior Commanders and Staff Officers at various levels! In my humble opinion, we should consider laid down Rules and Regulations as handrails to guide us and not handcuffs to tie us down. This maxim has to be applied diligently, unbiasedly and without any intention of covering up any foul play. In this very instant case, our taking little bit of risk in an unbiased and diligent manner has benefitted some needy dependents we never knew. Even now, I am very proud of having been a functional link in positive resolution of the unfortunate incident involving loss of the life of a bread winner of the family. May God bless all his family members.

# A MURDER MOST FOUL

In June 1987, I got posted out from my Battalion as DQ of an Independent Infantry Brigade Group located at Babina Cantt. Babina is rather non-descript a place close to Jhansi and not many people outside Armoured Corps circles are really familiar with the place. Infantry Brigade Group though located at Babina Cantt was not integral to the Armoured Division. Main Railway trunk line went right through Babina Cantt, dividing in to two halves. Infantry Brigade Group was located on the Eastern half, known as Old Cantt and all three Armoured Brigades on the Western half, known as New Cantt. Married quarters of all ranks were also generally in their respective areas.

On arrival at Babina Cantt, I noticed a strange practice in vogue. A vehicle mounted and armed mobile patrol used to be moving around in the Cantt area all 24 hours. Such Patrols may be a routine feature in insurgency affected areas but not in places like Babina where no such threat existed. It was revealed that a young Captain from an Armoured Regiment had been murdered one night by some unknown persons in New Cantt Area about a month ago. The movement of the mounted armed patrol was part of strengthening safety and security in the entire Cantt area.

Deceased officer was a tall, smart and newly married Sikh officer serving in a Regiment located in New Cantt area. He was a 2nd generation officer in the Regiment which his father commanded more than two decades ago. It was a common sight to see this couple happily speeding around on their Royal Enfield Bullet Motor Cycle in the Cantt. At times his wife could also be seen alone whizzing past on this Motor Cycle. Unfortunately, this very officer got murdered by some unknown persons in his house. A Station Court Of Inquiry was ordered to investigate the reasons for the occurrence of the crime as also to identify the persons involved in the heinous crime. Months passed but the Inquiry was not able to come to any conclusion, though by then it was already an open secret who all were the persons involved and how the crime was committed. Only the Members of the Court of Inquiry seem to be unaware.

Stories generally in circulation were that in the middle of the night of murder, Captain woke up to noise in his house and found three males and his wife in the adjoining Bedroom. It was not possible for one or even two persons to overpower the young Captain. Eventually, he got killed in the fight. It took few more hours before the scene of crime could be cleaned up by the assailants, hue and cry raised subsequently and matter reported.

Now the 'floor management' activities started. Father of the Captain was a highly regimental soldier and did not pursue the case any further for the sake of prestige of his Regiment. Late Captain's father-in-law was also a serving Army Officer and did not pursue the case for some unknown reasons. The end result was that life of a young officer was snuffed out for no fault of his and the culprits remained unpunished. But not really; definitely not all of them. One out of the three suspects in the crime got drowned in a nearby lake few months later while on a Picnic trip organised by the Division HQ. Divine justice seems to have prevailed to an extent.

Now, a word about the Senior Commanders who were in the chain of Command during and soon after the occurrence of this incidence and the level in Army hierarchy each one of them finally reached. Commanding Officer retired as a Major General. Brigade Commander and Division Commander became Lt Gens. Only person who suffered was the young Captain who gave his life for no fault of his and culprits remained unpunished.

# A NIGHT IN THE CREMATION GROUND

During 1984, our Battalion was located at Roorkee. We used to be participating in various training exercises in deserts of Rajasthan and canals in and around Roorkee. Once we were told to carry out a Brigade level 'Advance to Contact' exercise. We were to move out from Roorkee, mounted in vehicles, go Westwards for 50 – 60 kms to reach the base of lower Shivalik Ranges by midnight where the exercise was to be given a short tactical pause.

Everything went as planned. My Rifle Company was leading the advance. We went past many villages, towns, canals and finally reached the foothills of Shivalik Ranges and did our tactical maneuvering en route as required. By then, it was well past mid night and a pause of few hours in the training exercise was announced. Fortunately, my Company found a small open area in the jungles and we decided to halt there only. My Sahayak brought my Camp Cot from the administration vehicle which was following shortly behind, spread it in the open area and made requisite arrangements for me to have a short nap of 2 – 3 hours as training exercise was to resume shortly next day morning. Before hitting the Camp Cot, I noticed there was some fire burning 4 – 5 meters away from me and smoke also emanated from that place. It was bit nauseating. It did not bother me much and I soon fell to sleep.

Early morning next day, my Sahayak woke me up with a hot mug of tea. Getting a mug of tea was such a privilege at that time and place in the wilderness. I soon noticed the fire nearby was still burning though not that much what it was when we arrived there few hours ago. With a mug of tea inside, it soon dawned on me that we had spent night in a Cremation Ground of the nearby village and the fire burning the whole night was of the dead body brought by the villagers the previous evening and set on fire for cremation. Since the body was not fully burnt by the fall of night, the villagers had left the body still burning and returned to the village. In fact, we, unknowingly and unintentionally, had intruded in the otherwise quiet and sorrowful completion of the cremation ritual.

I soon got up from my Camp Cot, said my prayers to the God Almighty for grant of SADGATI to the departed soul of the deceased person. I suppose we, the Military persons, are exempted by the Almighty for such like unintentional intrusions during the course of performance of our Military duties.

# PART – II

# COMMAND AND STAFF

# AN AVOIDABLE TUSSLE

I had been working as GSO1 (Operations) at a Division HQ for nearly a year and half and waiting to take over command of the Battalion in which I had been commissioned. One fine day, in end Feb 91, I was surprised to see my posting order. I was required to take over command of another Battalion of my Regiment in a month's time which was scheduled to de-induct from its Formation shortly. I represented to Colonel of the Regiment but of no avail. I took over command of the Battalion in the first week of April 1991.

The battalion was scheduled to move out in 01 week of June and barely two months were left for its departure. Preparations for move out were already afoot, heavy baggage, other stores duly packed and necessary leave planning of Officers and JCOs put under implementation. I had known Brigade Commander and Staff at Brigade HQ and interacted with them regularly while serving at Division HQ. All appeared happy on my posting and looking forward to serving together. Within a fortnight of my moving in as CO, I realised they do not appreciate my way of commanding the Battalion. Frictions gave rise to sparks and kept escalating further. I was surprised by the turn of the events and could not fathom where I had faltered. My personal meetings with Brigade Commander on causal issues bore no fruits. It appeared he had blind faith on his Staff.

In mid-May, our preparations for de-induction were in full swing. Brigade HQ surprised us by sending an Annual Inspection programme lasting for one full week. Inspection events included Battalion Ceremonial Drill personally led by the Commanding Officer, Battalion BPET and PPT, Classification Firing, Platoon Battle Drill, Rock Climbing, visit to Battalion Quarter Guard, Stores, Regimental Institutes and a Formal Dinner Night. Discussions between Brigade Commander and Commanding Officer were scheduled on the sixth and final day. I perused the list of events we were to be put through just 20 days prior to our departure. It involved avoidable extra unpacking and re-packing of almost all Stores as also some of the officers had already proceeded on leave as per earlier implemented leave plan and were to rejoin duty at our next destination only. The aim of Brigade HQ appeared quite obvious.

Battalion went through all the events of the inspection and did well. We had a good set of Officers, JCOs and Senior NCOs and all of them had put in their best ensuring week-long inspection is a success. Sub Maj Saheb was a great asset. He had total control over all JCOs / OR and his contribution in overall functioning was no less than that of a Major. Brigade Commander could not find any major faults yet spoke sarcastically, sometimes even in the presence my junior Officers. Hints to avoid such comments were not heeded. A stage came when I decided 'this much and no more'. Before departing for final day's schedule of inspection, I told my wife, "In case I do not return home by 2 pm, we will leave for our village". We had 16 years of married life by then and she understood me well. She smiled and nodded in affirmation.

Commander visited all Regimental Institutes, Stores and Quarter Guard. Everything was well laid out. We reached CO's Office and I asked Staff Officers accompanying him to leave the CO's Office as they were not required to be privy to the discussion between Brigade Commander and Commanding Officer. Discussions started and he came out with numerous picayune and rather irrelevant observations on training and administration and lambasted me for no reasons. I asked Adjutant to send me a specific Register. I told him that his house-hold had not been making honest transactions with Brigade Shop run by the Battalion for quite some time. Only a Token Payment had been made against each transaction. All purchases and payments had been made by the Staff employed in Flag Staff House and mostly in the presence of wife of the Brigade Commander who mostly accompanied them. I put up the Register which contained the record of last two months transactions before him to peruse.

Initially, he tried to brow beat me. I stood firm on my stance. We two remained closeted in the CO's Office for more than five hours. He tried to wriggle out, saying he was not aware of this. I clearly maintained since it was happening in his house and for that long, he ought to have known it. Finally, I told him to report the matter to higher HQ for ordering an Inquiry against me for making false allegations. We left the office at 5.30 pm. No Officer or JCO had lunch till then as we two were still discussing things in the office. I did not hear of any Inquiry ordered by the higher HQ on the subject. After a fortnight or so, Battalion moved out from the Formation. Brigade Commander also went out on posting almost simultaneously. We honoured his request for transportation of his household baggage in our Military Special train and delivered at NRS to his place of posting.

A word about few of the officers who served with me those days in the Battalion. One Company Commander became COAS, two Major Generals, two Brigadiers and five Colonels. I am so thankful to the God Almighty for having blessed me with such a superb group of Officers, Sub Major, JCOs and Senior NCOs.

Such is the journey called life. I have thoroughly enjoyed my life in Army. Being a Graduate in pure Science subjects, I was offered to join any technical stream out of Engineers, Signals and EME. My choice was only INFANTRY and I stuck to that because this is where my heart was. I have endeavoured to follow a righteous path as guided by my INNERSELF. I well knew such a path was always full of challenges and pitfalls. I made a fair attempt to follow the basic values of life inherited from my family and learnt from numerous Senior Officers and Colleagues in due course of journey in Army life. God gave me opportunities to serve under truly professional and understanding Senior Officers. My colleagues always did a tremendous job and achieved the impossible. I am forever proud of my association with all of them.

# A BOXER'S DRUNKEN BRAWL

It was sometime in October/November 1991. Battalion had moved to area North of TAWANG in Arunachal Pradesh. Battalion HQ was located at a placed called NAKYO GG close to BUMLA. It was DASSAIN (DUSHEHRA) time. An entertainment programme, an essential part of the DASSAIN celebrations in a GORKHA Battalion, was in progress in a makeshift hall. It was around 9 pm and we all were enjoying the programme. Suddenly, duty JCO walked in, uttered some PHUS-PHUS in the ears of Sub Maj, who in turn conveyed same to Second-in-Command and Adjutant. All three, one by one, walked out from the entertainment programme and returned after 15 minutes or so. Soon after, I too was informed as to what had happened.

In fact, quite an unexpected incident had happened in the Battalion. Around 8.30 pm, Nb/Sub Adjt of the Battalion had gone to check the security of Battalion Quarter Guard. To his dismay, he found Sentry on duty drunk of liquor. Nb/Sub Adjutant, very rightly so, admonished the Sentry for this careless act while on Guard Duty at such an important place. Rifleman Partha Bahadur Thapa, a Boxer of some repute, was the Sentry on Quarter Guard duty. On being admonished, Partha Bahadur Thapa gave a furious hit on the face of the JCO thus breaking his jaw instantaneously which required early evacuation of the JCO to Field Ambulance at Tawang.

We took stock of the situation. JCO was immediately evacuated to Field Ambulance at Tawang. Offender was duly arrested. It was sad thing to happen and unpardonable by all means. If not dealt with sternly and expeditiously, it will breed lawlessness in the Battalion. If today it was Nb/Sub Adjt, tomorrow it could be Sub Maj or anyone else in the Battalion. Hence, I decided to send a correct message down the chain to all ranks that such acts of indiscipline will not be accepted. Accordingly, a Court of Inquiry and followed by a Summary of Evidence were ordered. Both were duly completed by next day morning. Offender was duly tried, dismissed from Service, given one year's imprisonment and handed over to Civil Jail the very next day.

It was definitely a sad thing to happen. Equally sad was to dismiss a person from service running in his final year of Service to earn his full Pension. Dismissal from Service denied him Pension and all other related benefits. However, gravity of Offence left no alternative before me but to give an exemplary punishment. I never regretted doing it. It was later learnt that Partha Bahadur had indulged in such manhandling of his colleagues, even seniors, earlier too when under the influence of liquor. However, he was never punished which may have encouraged him to continue doing so.

I commanded Battalion for almost three years and moved out on posting. Much later to this unfortunate incident, I was posted as Col 'Q' in Andhra Sub Area. Sometime in mid-1998, Officer-in-Charge, Legal Cell walked in my office. He had to submit comments to Andhra Pradesh High Court on the case of a Gorkha Soldier who had been dismissed from Service. Since my name had figured in the case as a Commanding Officer of the accused, he asked if I would like say something. I told him that the accused person had been punished for a grave offence and the circumstances under which he committed the offence deserved an exemplary punishment. Whether High Court now decides to commute or cancel the punishment or grant pensionary benefits to him, I have nothing to say and it was for the High Court to decide. After few months, I moved on posting to Srinagar (J &K) and could not keep track of the case whether he got any remission in the punishment. May God grant SADBUDDHI to Rifleman Partha Bahadur Thapa and offer my sincere good wishes to his family members wherever they are.

# AN ADVENTURE TRIP THAT ALMOST TRIPPED

While commanding the Battalion in High Altitude Area, I had the good fortune of having a number of junior officers from various Services ie ASC, AOC, EME and AMC attached with the Battalion for six months to one year. In January 1992, 2/Lt Neeraj Dobhal, an officer from ASC, commissioned from Officers' Training Academy was attached with the Battalion for a period of six months.

Neeraj Dobhal was son of an Air Force person and belonged to Dehra Dun. He had lost both his parents in his childhood itself in a scooter accident and he and his younger sister were brought up by their maternal grandparents all through. I had little soft corner for him owing to this background and wanted him to complete his tenure in High Altitude Area safely.

We had a forward Post at an altitude of approx 15,000 ft plus, right on the Border and overlooking deep into enemy territory. Its approximate 300 meters long final approach was very steep and slippery, making it an interesting but arduous climb, especially during rains and prolonged winters with heavy snow accumulated all along. Neeraj Dobhal wanted to spend at least one month of his attachment period on this Post as he will never get an opportunity to live under such terrain and climatic conditions in his Army career again. An opportunity came calling soon when, in April 1992, the Post Commander went on a short leave and Neeraj Dobhal sent as his relief.

It was routine drill for all Post Commanders to report to the CO personally at 6 pm every evening and brief on all important activities carried out at Post during the day. However, one day, a Havildar from this Post gave the evening Report to me and not the Post Commander or even the Post JCO. He further told that 2/Lt Saheb and Post JCO both had gone down and not returned as yet. I rang up again after 30 minutes and was given the same reply. It was already dark and night temperature in those areas plummets to minus 10 degrees as a routine. I soon realised we have a problem at hand tonight.

There was a Hot Water Spring approximately7 - 8 km downhill which Neeraj Dobhal wanted to visit. He discussed this with the JCO and accompanied by three more Soldiers from the Post walked down to the area of Hot Spring. They somehow got late in their return journey. Also, there was 2 to 3 ft of fresh and accumulated snow along major part of the route. Going downhill was enjoyable but climbing up was quite difficult. Midway, Neeraj Dobhal got tired and stage came when was literally unable to walk any further. Remainder Party was not equipped to carry him on their shoulders as there was steep ascent and thick snow. They found a big rock en route, put Neeraj Dobhal under that Rock and moved ahead to ask for assistance to evacuate him.

Meanwhile, the troops from the Company deployed in the near vicinity had been fully mobilised and searching for the missing persons. We well knew that, under these adverse climatic conditions, every minute matters in saving a life. A person not saved in time or, as in this case - left under a rock, will soon get frozen to death. At around 10.30 pm, the JCO and his Party of three Soldiers made contact with the Search Parties and gave information about the whereabouts of the officer left behind. By mid-night, he was rescued and brought to the Company HQ. He was found in a bad physical and mental state when located sitting under the rock all alone and under sub-zero temperature conditions. I heaved a sigh of relief when he was safely brought up to Company HQ. I thanked God Almighty for saving the life of this officer who had lost both his parents in the childhood.

2/Lt Neeraj Dobhal was brought to the Battalion HQ after few days. A very simple soul, gone through a difficult time just few days ago. I asked him whether he will like to visit Hot Springs again. He immediately sat straight in his chair and replied in a big affirmation. That was truly a great reply to hear from a young officer. I felt very relieved and contented when he was finally dined out from our Battalion on successful completion of his attachment. He was posted to an ASC Unit near Jorhat. After a year or so, he requested for his posting to Dehra Dun to facilitate his younger sister's marriage. This somehow worked out well and he was able to perform his responsibility toward his sister.

I retired in 2003 and had no connection with him for almost two decades. I had a habit of noting down permanent home addresses of all my Battalion and attached officers. I was working for Hotel Savoy, Mussoorie in 2011-13 and tried to contact Neeraj on his Dehra Dun address. That house had long been sold out but I could trace his relations and we were in contact with each other again. It was a great feeling to speak to him after such a long time. He left Army after five years, joined Security Division of Reserve Bank of India and did well there too. May God bless you, dear Neeraj and all your family members, always and everywhere.

# MY IMPRESSIONS OF CHINESE TROOPS

I was commanding Battalion in Area North of Tawang in Kameng Sector of Arunachal Pradesh in 1991 - 93. It was my second tenure in the same area within four years. Our Battalion had spread from LANDA GG in the East to SULULA in the West. SULULA had an interestingly arduous climb starting from SHUNGATSAR lake and was located on Eastern flank of WANGDUNG / SUMDRO RONG Valley. In fact, famous SUMDRORONG NALA emanated from general area SULULA itself.

WANDUNG / SUMDRO RONG Valley had been quite a flash point in Sino-Indian relations for almost a decade during 1986 to 1995. It was during the summer of 1986 that Chinese Army quietly occupied this Valley, thus creating lot of thunder in Sino-Indian relations till it vacated at its own in mid 1990s. I was serving with my previous Battalion as a Company Commander in that area during those days, had an opportunity to visit SULULA and LUNGLROLA few times and observe activities of Chinese troops in the course of my duty. I had observed that Chinese troops deployed in WANGDUNG Valley were neither physically strong nor well trained. They did not follow correct battle drills and even were afraid to operate in small groups at night.

Now in 1991, I was back again in the same area, this time as Commanding Officer of an Infantry Battalion. We interacted with the Chinese troops at SULULA and BUMLA bit more frequently. Some interactions were formal and were conducted in the form of six-monthly FLAG MEETINGS at BUMLA Pass. We had numerous unofficial interactions with Chinese troops so as to resolve certain local issues at our own level without referring to higher HQ and held on as required basis. Since our troops at SULULA occupied a much higher ground with just a distance of 500 meters separating us, we were able to monitor every activity of the adversary in

WANGDUNG VALLEY closely. I found no change to my observations made four years ago. Chinese were still not well trained. They were neither properly motivated nor physically fit to operate in the area occupied by them. Chinese troops were not deployed face to face opposite BUMLA Pass area and were 5 – 6 km inside their territory. Hence, our interactions with them in this area were generally restricted to six monthly FLAG MEETINGS or informal meetings at local levels on as required basis. Just to quote an interesting incident connected with it. One day, sometime in end 1992, I got a call from Brigade Commander that the Div Cdr is at Tawang and will be visiting BUMLA next day and made a passing reference if some persons from opposite side could also be there. Likely arrival of the GOC was not to be disclosed to many people in the Battalion. GOC was surprised to see Chinese troops at the Pass and asked me how did I manage it. My answer was, "Sir, it just so happened".

Sometime in September 1991, I asked our Company Commander at SULULA, to organise an informal meeting with the Commander of Chinese troops deployed in WANGDUNG Valley. This was to be an informal meeting held at local level and no information was sent to our higher HQ. My intention was to avoid escalation of minor activities to an international level where troops intruding each other's territory mistakenly are exchanged through New Delhi or Beijing. I wanted to put my own mechanism in place to implement it, even if a risk has to be taken to start with.

Accordingly, a place downstream in SUMDRO RONG was selected as venue, date and time fixed and Chinese Commander informed. Both the parties arrived, in fact Chinese Company Commander and his soldiers were waiting for our arrival. Chinese Party consisted of three officers and seven soldiers. We were four officers, including self. Chinese Party had an Interpreter with them who could converse in workable English and Hindi. This interpreter, subsequently was to become a frequent common link in all our formal or informal meetings held during remainder part of our tenure in that area. We spent more than an hour together. We shared our breakfast with each other, took photos; few Chinese soldiers even interchanged their headgears with us for a while and took photos. It was so difficult to differentiate between a Gorkha and a Chinese soldier with headgears interchanged.

Suddenly, an Idea came to my mind: why not use this bonhomie in visiting the Chinese Post at WANGDUNG. Though the risk involved was huge. Also, it had not been discussed with any one even in my own Battalion HQ nor with the officers at SULULA Post who were accompanying me for this meeting. I was of the opinion that there is no harm in it and the risk is worth taking. Before breaking off, I spoke to the Chinese Commander and told him that I want to visit his Post at WANGDUNG and have a cup of Tea with him there right now. Chinese Post Commander immediately got up and respectfully told me to come and have a cup of tea or a meal with him right now. We both started moving down towards WANGDUNG Post. SULULA Company Commander and remainder officers were listening to our conversation and were rather surprised at the turn of events. Two of them got hold of my Coat Parka from behind and spoke in GORKHALI," HUZOOR, AJU KO LAGI TOH AITI NI HUNCHHHA, AGADI JANU PARDAINA " ( Sir, this much is sufficient for today. We need not go ahead any further ). I took an instant decision. I changed my decision in a flash of moment and conveyed the same to my Chinese counterpart and made a promise of a visit to his Post in WANGDUNG next time. After few months, I confided this to our Brigade Commander at TAWANG. He had a hearty laugh and said, "R V, TU KISI DIN MUJHE GHAR BHIJWA DEGA".

Even today, I think I should not have changed my mind at the last moment. I should have gone ahead to visit the Chinese Post at WANGDUNG, enjoyed the hospitality of my Counterpart, and may be, that I then returned to India via BEIJING.

# A CORPS WARGAME

Battalion was deployed in a High Altitude Area in Arunachal Pradesh. There used to be heavy snowfall from October to April every year. Hence, all movements ahead of Battalion HQ were restricted to 'on-foot' only. My bi-monthly visits to all forward posts made me fully acquainted with the terrain Battalion was occupying as well as area as opposite it. I even made a few forays into the enemy territory opposite my area and Brigade Commander kept in the know of. This gave me a detailed insight into the various options available to the adversary opposite my battalion area and the restrictions terrain may impose upon him if he decides to play any mischief.

In Feb/March 1992, I was detailed to participate in a Corps Wargame as part of 'Enemy Syndicate'. Enemy Syndicate projected a plan in which a Brigade-size force was to launch an attack along an axis that was difficult but not impossible. 'Defending Syndicate' led by GOC of our own Division vehemently opposed it. A stage came when 'Defending Syndicate' and 'Enemy Syndicate' both did not buzz from their plans. 'Enemy Syndicate' plan had been based on terrain information as provided by me which facilitated movement and subsequent sustenance of a Brigade-size force along this axis. Resultantly, Corps Wargame came to a standstill and it required an intervention of the Corps Commander. Corps Commander once again discussed the Enemy Syndicate Plan and approved it. However, he asked me to give presentation on terrain as existing opposite the Divisional Defended Sector to all participants of the Wargame. This was with an aim to convince all attendees of the feasibility and subsequent sustenance of the Brigade size force on that axis. After presentation, the Corps Commander was fully convinced, approved the 'Enemy Syndicate' plan and Corps Wargame progressed accordingly. I did not realise that our Division Commander had taken this small thing to his heart and will take it out on me some day later.

Our Brigade Commander was a fine gentleman and a true ground soldier. He had seen our Battalion for last one year and subsequently initiated an 'OUTSTANDING' performance report on me. This was my first ( and last too ) such Report in my entire Army career spanning 35 years. Obviously, I felt happy. Alas! This happiness was rather short-lived. After two months I was called to Brigade HQ at Tawang for an important meeting with the Brigade Commander. In actual fact, Division Commander had returned my performance report bringing out a technical flaw of Division HQ not been informed three months prior to initiation of such a Report. I left it to the Brigade Commander to resolve the way he wanted and the same acceptable to me. He altered the overall grading from 'OUTSTANDING' to one grade lower on the same Report with no other alterations made elsewhere. We both signed on the alterations made. I forgot about this as things were beyond my control and left everything to God Almighty.

Meanwhile, Corps Commander got posted out as Army Commander in the same theatre. Few months later, one fine day while sitting in my Battalion HQ at Bumla, I got a telephone call from Command HQ. Person other side introduced himself as MA to Army Commander. He said that he was speaking to me on directions from Army Commander who has conveyed I need not worry about my Performance Report and aberrations had been set aside. Entire conversation lasted barely for 2 minutes, I could not even thank him.

Army Commander was a very fine soldier. He was our DS in DSSC, Wellington, when I attended Staff Course in 1981. Just a few months later, this great soldier, along with his wife and Col MA, died in a helicopter crash. What a tragedy! A great loss, indeed. Such fine soldiers are not born every day. May God grant peace to the departed souls.

I always had great faith in the ARMY as an organisation. Also, I have experienced that if a person has basic character qualities, competence and a correct belief system, one will surely find support from somewhere which would rescue him in the hour of need. As stated earlier too, I never got an 'OUTSTANDING' report in my entire career yet was never superseded till my retirement. Army is a GREAT organisation. I never had any doubts about it. "God has an overall plan. We all have to play our part in it. So play our part well" has rightly been said by some wise man.

# POSTING OF A GORKHA REGIMENT OFFICER AS COMMANDANT, OTA, CHENNAI

Our Army is a great monolith of an organization. Sometimes it functions in a manner that may appear strange and abrupt at first glance, yet it proves to have a lot of experience and intellect behind such prompt decision-making.

I was commanding my Battalion 4/5 Gorkha Rifles at Chennai for a few months during end 1993. Major General V Rajaram, 4 GR, was the Commandant, Officers' Training Academy, Chennai. I happened to serve under him earlier as an Instructor Class 'B' (Major) in 1981-82 when he, as a Lt Col, was Commander, Young Officers Wing, Infantry School, Belgaum. May God bless his soul as he died at Secunderabad few years ago. He was a great Commander, and I always had immense regards for him. During one of our interactions with him at his residence in Chennai, he narrated an interesting circumstances related to his posting as Commandant of the Academy.

General Bipin Joshi, then COAS, visited Academy during the tenure of predecessor of Maj Gen V Rajaram as Commandant. Previous Commandant complained to the COAS that a disproportionate number of officers from various Gorkha Regiments had been posted to the Academy. In that, he brought out that both the Training Battalion Commanders and Adjutant, few Captains as Instructor Class 'C' are all from Gorkha Regiments, leave aside a Training Company from 3 and 9 GR. He asked COAS if this imbalance could be rectified.

Lo and behold! In no time, the previous Commandant was posted out and a Gorkha Regiment Officer (Major General V Rajaram in this instant case) posted in his place. What a neat and clean arrangement; created the least possible turbulence in the environment and achieved aim fully. That we may call as quick decision making with lot of intellect behind. Made everyone happy, including both outgoing and incoming Commandants.

# DAL - BHAT EATING LAMB FROM BUMLA

It was sometime in March / April 1993. Battalion was located at Bumla, North of Tawang in Kameng Sector of Arunachal Pradesh. We were nearing completion of our two years long High Altitude Area tenure and designated to go to Chennai. What a change from minus 15 degrees to plus 50 degrees of temperatures! It was quite normal at Bumla to have heavy snow fall till May and two to three feet of accumulated snow. Preparations were afoot to pack all heavy stores and shift them to our rear echelon at Tawang. Myself and Sub Maj Saheb used to take a periodic round of our Stores, FRP (Fabricated Reinforced Plastic) Huts and Tin Sheds and take a stock of on-going progress made so far.

It happened on one such day when we two were taking a round of the Battalion area and entered in Tin Shed where some of our Vehicle drivers and other administration staff were living. Tin shed accommodation had double decker Cots and was suitably insulated from inside to beat freezing cold outside. Moment we entered inside the Shed, it smelled as if we have entered in some Animal Shed. Also, we found some animal droppings scattered on the floor. Suddenly, Sub Maj Saheb noticed a small 5 - 6 months old lamb tied to one of the Cot and, seeing all of us in his abode, started continuously bleating. A soldier present there told us the lamb belongs to one of the Vehicle drivers who found it abandoned on the road few months ago prior to onset of last Winter Season.

Sub Maj Saheb investigated the matter and an interesting story was revealed. Locals from Tawang and numerous nearby villages have divided all hills North of Tawang amongst various graziers for grazing of their animals i.e. sheep, goats, Yaks and horses. This is a wisely devised system based on mutual understanding and honestly followed. All graziers from Tawang area take their herds to respective areas in the month of April / May and return in October

/ November each year. During November 1992, a 1 Ton vehicle was returning from a Forward Post to Battalion HQ at KLEMTA and, at the same time, a grazier was also taking his sheep herd back from forward areas to Tawang along the same road. It was during this move that a sheep delivered a baby lamb. Grazier, somehow, did not notice it and, in the rush, mother sheep also could not halt for long to tend to the newly born lamb. Resultantly, the poor lamb got left behind. This lamb was destined to survive as the Driver of the Vehicle returning from Forward Post noticed the lamb lying in the middle of the track. He felt pity on the lamb and decided to keep it with him in his living area. Not a blade of grass grows North of Tawang from November to April each year. How did the lamb survive without its natural feed and against extreme cold weather! Driver could only feed the lamb what was available in the Company kitchen; that was POORI- SABZI in Breakfast and DAL- BHAT in Lunch and Dinner. Lamb got used to eating what so ever was cooked in the Company Kitchen, grew quite healthy and active and was loved by all. Now it got certification from CO and Sub Maj also.

But the poor lamb was not destined to survive for long. Battalion moved down to Tawang in middle of June 1993 and halted for a week. Lamb was also brought down for its onward journey to Chennai. One day it was left free for grazing near a Nala where some grass had grown. It did not recognize edible from non-edible grass and ate what so ever came in its way. It soon fell sick and died. It was a small tragedy for the poor Driver and something to mourn for many others too.

*A Paltan BARAKHANA at Dharamsala (HP), 1991.*

*Beautiful Winters of 1991 - 93.*

*Beautiful Winters of 1991 - 93.*

*With some Chinese Soldiers - 1992.*

# PROMULGATION OF A COURT-MARTIAL PUNISHMENT

It happened sometime in 1994. I was posted as Colonel Administration of a Division at Bareilly. There had been a case in which an Officer from JAG Branch posted to nearby Area HQ, had been caught accepting bribe from a Recruiting Medical Officer of a different Military Station. Medical Officer himself had been through a legal proceeding for accepting bribe from otherwise medically unfit candidates and was awaiting final disposal of the case from Higher HQ. At this stage, this Major from JAG Branch tried to exploit the Medical Officer who in turn informed Higher HQ. Accordingly, a trap was laid and the JAG officer caught red handed by a group of officers; an act almost similar to any Bollywood Thriller.

Case had been finalised and JAG Officer was to be 'Cashiered out from Service' as punishment for the crime committed. He had been attached with our Division HQ pending confirmation of the Sentence which was soon confirmed. Punishment awarded had to be now promulgated by the Divisional Commander. The date of promulgation arrived and the Officer was brought to the Division HQ. He wore his Major rank badges on his shoulders in cloth epaulettes for ease of removal during the promulgation. He was duly marched to the GOC, punishment read out and all legal formalities duly completed. GOC removed epaulette from one shoulder of the officer and asked me to remove the epaulette from other shoulder. We all have gone through the joyful occasion of our own 'Pipping Ceremony' or that of our wards. Removal of other person's rank badges, irrespective of the gravity of crime committed, does make a person feel sorrowful. It does shake one's inner self, even if for few seconds. After all a person standing before you was a Major in Army few minutes ago, now he is nothing and has lost all privileges of being part of this elite organisation. Yes, he had definitely erred. He is standing in front of you and is no body now. Before leaving the GOC Office, he requested if he could wear jacket over his shirt while leaving the Div HQ. He did not want to face any lower staff as now he had no rank badges on his shoulders. This was granted to him. Worst was that I had to leave him at his Guest Room where his wife had been waiting for some miracle to happen to save her husband. Moment he reached Guest Room, lot of crying and wailing took place. I was reminded of numerous such occasions which I had experienced in Nepal 20 years ago. During 1972, I visited the villages of number of persons from my Battalion who were martyred in 1971 Indo-Pak War. It used to be somewhat a similar scene of wailing and crying by the widows and parents of the deceased persons; only cause and purpose were different. Those martyred had placed name of their Battalion on a higher pedestal where as the case of this person was purely of self aggrandisement.

He appealed to the COAS for mercy and, after few months, his Mercy Plea was partially accepted. Punishment was downgraded from 'Cashiering Out' to 'Removal from Service' or words to that effect, thus enabling him to become eligible to apply for employment elsewhere too. Fair one, I suppose. He was too young in age; had to live a long life ahead, support his wife, children and may be even dependent parents too. An opportunity was provided to him to start a life afresh.

# PUZZLES OF ADMINISTRATION

I was Colonel Administration of a Division. I have been Staff Captain in a Brigade, later DQ in an Infantry Brigade Group and now doing Colonel Administration of a Division; this all was very interesting. Three Admin Staff appointments and fourth one as GSO1 (Operations) in a Division HQ, was a good going. It was natural for many to presume I was a trained administrator, which I never was. Some COs of Div Troops Units used to often seek my advice on various matters of administration as experienced by them in their Units. It was amazing to see how a human mind reacts to certain age old dogmas and beliefs prevailing in a Society irrespective of caste or creed or religion or place. Also, I saw how things go out of hand if not handled timely, firmly and with maturity.

## Curse of Narrow Societal Beliefs

A senior Major of a Div Troops Unit was facing a very strange problem in his married life for which neither he nor his wife was responsible. CO of the Unit tried his best to resolve but could not and now sought my help. I requested CO to inform the Officer that Colonel Administration and his wife have invited them for a cup of tea at their residence and, if consented by them, we all may discuss their marital issue too.

It was truly a strange case. Officer was from a well to do agriculturist family of Andhra Pradesh and married few years ago. On the face of it, both of them appeared to be happily married. Problem arose when Major's younger sister was married to the younger brother of his wife who was a serving Captain in AMC. Marriage took place with the mutual consent and beliefs of both the families. Soon after the marriage, all hell broke loose. When the girl first time visited her parents after marriage, she complained to her mother that her husband suffered from impotency. She further told that her husband fully well knew of his deficiency before marriage but did not reveal it to anyone. This became a serious issue between the two families and still could have been resolved in many peaceful ways by both the families. However, the parents of the girl wanted to take revenge and that too in a strange manner. They wanted the Major to divorce his wife who was the younger sister of the Captain. Her only fault was that she was the sister of the defaulter. In the whole episode, only person to be blamed was Captain who should not have got married to the girl fully well knowing of his deficiency. Major was unable to go against the directions of his parents. Though he loved his wife and never had any marital discord with her, still he was not able to stand up against the vengeful directions of his parents. His wife was in a terrible situation as she loved her husband but had become a mere pawn in the revenge game played between the two families. She was prepared to suffer the agony of a divorce if her husband insisted and did not want to go to the Court to fight her case as this will go against the reputation of her parents. We discussed the issue with both of them at great length but could not convince either of them. We broke off quite disheartened. What a cruel way of taking revenge!

After few months, I went on Higher Command Course and almost lost track of the case. I got posted as Col 'Q' of Andhra Sub Area, Secunderabad and happened to meet erstwhile Second-in-Command of the same Unit who was then on Study Leave. I enquired about the case and was told that the Major and his wife went through CONSENSUAL DIVORCE soon after. What a tragedy! So many happy lives disturbed just for the sake of a family revenge when there were numerous other options available to both the families. May God grant SADBUDDHI to all to be brave enough to wade through these age old shackles and make place for a better world.

## CLASH OF EGOS

It was a simple case of a clash of Titanic egos of two personalities and I had been briefed on the case in detail during my taking over from my predecessor. In fact, my predecessor also did not know much about this as he was Col Q till few days ago and Col A had gone on posting. I was to now head the merged A and Q Branches. May be it could have been better if handled timely and at an appropriate level before it assumed such a proportion.

It happened between the wife of a Brigade Commander and wife of CO of an Infantry Battalion. It appeared that both the ladies differed on content and methodology of conduct of certain Social events by the Battalion. Brigade Commander and Commanding Officer were from different Arms, hence, perceptional differences between the two ladies on conduct of such events. This had happened on more than one occasions; hence, tension kept building up slowly. Finally, a stage came when, one day, CO's wife asked Brigade Commander's wife to leave the venue. This became a serious issue and tensions rose high between Brigade commander and CO of the Infantry Battalion. Division HQ was monitoring the case. Even the GOC and his wife went to the Brigade. GOC had detailed discussions with Brigade Commander and CO of Infantry Battalion and GOC's wife with the two affected ladies and other ladies too. However, the case remained unresolved with lot of tension writ large between Brigade HQ and Infantry Battalion. A highly avoidable situation and anything could have happened; even a minor spark could have culminated in a major avoidable incident. It did not happen does not signify low probability.

GOC had completed his tenure was soon posted out. Infantry Battalion had almost completed its tenure and moved out to a Field Area after a few months. Brigade Commander and his wife were the only two participants of the above mentioned episode now left in that Station. Annual Confidential Reports of all Colonels were under process. One day, new GOC handed me the ACR of the CO of the same Infantry Battalion to process, saying, "See it, you may not get an opportunity to see such a report again in your Army Career". I was truly amazed. Brigade Commander had taken out his full revenge upon the CO. There could be nothing lower than that. Report was duly processed to Army HQ. Few months later, I got a call from Complaints Advisory Board, Army HQ. Colonel at the other end was an old friend and wanted me to brief him on the case of CO Infantry Battalion as he had put up a complaint against the ACR. I asked him whether he wanted a single line answer or a lengthy Staff Officer-like answer. He confirmed me to give him a single line answer. I just said, " Sir, it was a CLASH OF EGOS OF TWO SENIOR LADIES" and our transaction finished at that. ACR was duly set aside.

However, story had one more chapter to unfold. Brigade Commander completed his tenure and went on posting as Sub Area Commander. CO Infantry Battalion Commander also completed his command tenure in due course and received posting order as a Senior Staff Officer in the same Sub Area HQ. Fortunately, the posting order was cancelled and future unseemly occurrences avoided.

# SCARE IN THE AIR

It happened sometime in mid-1997. I was posted as Col 'Q' at HQ Andhra Sub Area, Secunderabad. I was required to accompany Sub Area Commander to Visakhapattanam for conduct of Annual Inspection of few Station Units. Indian Airlines, the lone Air Operator between Hyderabad and Visakhapatnam, had suspended its Flights more than a year ago for some technical reasons and was to resume it connectivity soon. We both travelled by train from Secunderabad to Visakhapatnam. However, we did manage to get two seats in the inaugural Flight of Indian Airlines for our return journey to Hyderabad.

We reached Visakhapatnam by train next day morning and carried out the Inspection of Station Units, as scheduled. We still had lot of time left before catching return journey Air Flight in the afternoon next day. I utilised this time in visiting a Shiva temple built on a prominent hill North of Visakhapatnam. A large Shiva statue had been placed on a huge rock face which made it a prominent landmark and could be seen from far and wide. Also, one could have a beautiful view of entire Visakhapattanam town, Harbour and Naval Base in South, as also deep into Bay of Bengal in the East.

Indian Airlines had made lot of publicity regarding resumption of Flights and one could see lot of bonhomie at the Air Terminal. Visakhapatnam does not have a separate Civil Air Terminal and makes use of Naval Air Base for all its Flight requirements. Flight had been booked to its full capacity. Numerous political and civil officials of prominence were on the Flight. Two sitting Members of Parliament representing the adjoining constituencies, CMD of Rashtriya Chemicals and Fertilisers, a senior lady IPS Officer posted in Police HQ at Hyderabad and we two were representing Army. There were other important persons from nearby areas who were also travelling by the same Flight to Hyderabad.

It was a twin engine Boeing 737 Aircraft and took off from Visakhapattanam at given time. Flight duration between the two Cities was just 45 minutes. We two were sitting on adjacent seats somewhere in the middle. After 15 minutes of being airborne, we noticed Aircraft had not gained sufficient height and was flying at almost tree top level. Soon we recollected there was an abnormal sound and bit of jerk felt in the Aircraft little while ago. We both cautiously looked outside through window; aircraft was still flying at rather low a level. There were many others like us sitting in the aircraft and looking strangely at each other without knowing what has actually happened. I nudged Sub Area Commander and told him something appears to have gone wrong. He also was well aware of it and kept composed. By now we had overshot our travel duration and were nowhere near Hyderabad. It was at this time that Captain of the Aircraft came on air and announced there had been a technical glitch with the Aircraft and now returning to Visakhapatnam.

Aircraft first headed North of Visakhapatnam. I could well recognise the Hill on which Shiva statue had been placed and visited by me the previous day. Aircraft then headed East, toward the Bay of Bengal. We all were worried but could do nothing. I could see faces all around me - men and women, young and old - all appeared much more worried than me. We then flew over dark blue waters of Bay of Bengal and after a while heard sound as if something has been dumped down from the Aircraft. In fact it was additional fuel which had been thrown by the Captain of the Aircraft in to the Sea. Aircraft retained barest minimum fuel sufficient enough for an emergency landing so as to minimise the damage in case it catches fire while landing. Soon the Aircraft touched the landing runway at the Naval Base and came to standstill within 200 to 300 meters of touch down. It was a very short and smoothly managed landing by the Pilot.

Moment Aircraft came to standstill, many passengers rushed to open the exit gates themselves to be the first to jump down. None of them was listening to requests made by the Cabin Crew to remain seated till ladders are placed. Flight Captain again came on air and admonished the unruly passengers that though the aircraft has safely landed but few of them will surely break their limbs by falling of the Aircraft without arrival of the ladders.

We both looked through the windows on to the Right Wing of the Aircraft and were left aghast. The Right Wing Engine had burst and was almost hanging from its socket. This was the reason our Aircraft could not pick up requisite height as it was flying with just one engine only. It was an amazing feat by the Flight Captain to land the Aircraft so safely and smoothly with just one engine operating. Once out from the Aircraft, we noticed entire Naval Aviation Base on highest alert for an emergency landing. Numerous Fire Tenders had been mustered from the City also to cater for any eventuality. We managed a seat for the Sub Area Commander in the same day train while I reached Secunderabad next day evening.

The story of crash landing of inaugural flight of Indian Airlines from Visakhapattanam to Hyderabad had broken into all leading Newspapers of South India next day. I had not informed my wife about this till I reached home. My son who was on holidays from Pune, read it in the Newspaper and hid it from my wife. Many telephone calls from well wishers on my residential landline telephone - mobiles were still a far cry those days - broke the news to my wife and she remained worried till I reached home in the evening.

It was a very close call and anything could have happened that day. A great experience to undergo when one is unable to do anything to thwart it; other than silently praying to God Almighty, saying AB CHHOD DIYA IS JEEVAN KA SAB BHAR TUMHARE HATHON MEIN, HAI JEET TUMHARE HATHON MEIN AUR HAR TUMHARE HATHON MEIN. Our destiny is in the hands of someone else, the God Almighty; better leave it to Him to manage it.

# COURT-MARTIAL OF AN OFFICER FROM DGQA

It was sometime in the year 2001. I was commanding a Brigade at Kanpur. Kanpur is an old Industrial hub in Uttar Pradesh and houses numerous private Industries established during Pre and Post Independence era. A large number of Ordnance Factories were also established in the City during the previous Century. A requisite Quality Control Assurance (DGQA) set up had also been established alongside at Kanpur for close coordination with the Staff of Ordnance Factory Board. Members of DGQA are personnel from Armed Forces but seconded to DGQA to carry out supervision of Quality Control of the items produced by the Ordnance Factories as well as those bought from local resources. A fair number of them belong to EME, AOC and few from other Arms and Services too. DGQA has a well laid out system of having detailed instructions on each every aspect an item required to be produced by the Ordnance Factories or procured from local producers. These instructions are exhaustive, covers each and every aspect a final product should have. If a quality Assurance Team has done its job properly, there is no reason a sub standard or poor quality item could ever get introduced in the Armed Forces. Still, we often come across supply of poor quality of items supplied by Ordnance Factories or procured indigenously. Reason is quite obvious; numerous persons in Quality Assurance get compromised.

It was in the year 2001. I was informed that the Court Martial of a Brigadier from DGQA is to assemble in the location of our Brigade at Kanpur. Presiding Officer was a Maj Gen and four Brigadiers were Members of the Court Martial. I was also a Member of the Court. Court of Inquiry and Summary of Evidence had been completed and the Officer blamed for the serious irregularities committed by him. In fact this was an old case, detected almost over four years ago and accused Brigadier had been under close

Military Custody ever since. Legal process to prosecute the officer became very lengthy and had now reached the final stage of Court Martial. It took us three months to deliberate on each and every aspect of existing laws, Army Act / DGQA Instructions and his culpability thereon. I felt proud of existence of such detailed Instructions even on a minute aspect of Quality of an item, yet it was equally disappointing to learn how blatantly and cleverly people had been cheating the system and getting away with it. Brigadier under fire was a senior officer in the system and had been found to be indulging in such heinous acts for long and without any sense of guilt. He tried his level best to get away but could not and Court awarded the punishment of 'Cashiering Out from Service' with five years of Rigorous Imprisonment in Civil Jail. There was much weeping and wailing by his wife and other family members when the decision of the Court was pronounced. It may be noted that it is always the wife and children of the accused who are main beneficiaries of the ill gotten wealth. Officer was again sent to Military Custody pending confirmation of the award of the punishment by the Competent Authority.

Confirmation of the award of the punishment by the Competent Authority took few more months. By then the accused had completed five years in close Military Custody. It was considered equivalent to being in Civil Jail and the accused went straight to his home in Chandigarh without spending even a single day in Civil Jail. What a travesty of justice ! Shame to our Legal System !! Wish it could bring in right sense of expediency in itself. I remember of what Justice J C Shah, who investigated excesses committed during Emergency of 1975, had said. He said, "JUSTICE SHOULD NOT ONLY BE DONE, IT SHOULD ALSO SEEM TO HAVE BEEN DONE". Court pronounced the accused a 'Five years Rigorous Imprisonment in Civil Jail' but could not send him to jail even for a day as he manipulated the system and ensured that his trial is unduly prolonged and even confirmation of the sentence is delayed appropriately so as to avoid his going to Civil Jail. This is our dear DEMOCRACY; high and mighty get away with murder and poor languish in jail unheard.

# DEEP POCKETS

It happened sometime in the year 2000-01. I was commanding a Brigade in Kanpur. A major corruption scandal broke out in Army HQ / Ministry of Defense involving numerous senior officials dealing with the subject of procurement of Weapons and Equipment for the Defense Forces from foreign countries as well as manufactured indigenously. Names of serving Major Generals involved in the case, were hogging the limelight in Print and Electronic Media for a long time. Comment by one of the two accused officers as "Requirement of having deep pockets" was shameful and brought a bad name for the defence establishment in general and Army in particular.

It is often experienced that higher HQ tends to take a lenient view of the cases of moral turpitude involving Brigadiers and above and deals sternly in cases of Colonels and below. The similar outcome was expected in this case too and the defaulting senior officers expected to be let off lightly for some weird reason or other.

I decided to write a D O Letter on this matter to General S Padmanabhan, then Chief of Army Staff directly. I wrote a short, respectful yet strongly worded letter to the Chief, asking him to award exemplary punishment to the defaulters and no REGIMENTAL OR SCHOOL BOY AFFILIATION should come in the way of dispensation of the justice. It may be noted that one of the two accused Maj Gen was not only from Chief's Regiment but RIMCOLIAN too, hence, award of a lighter punishment to him was very much expected. Since it was a Demi-Official Letter, I neither sought prior permission of any superior HQ nor was any one in chain of command informed of my action. Letter was sent to the Secretariat of the COAS by Registered SDS to ensure its delivery. I neither received any reply nor was any acknowledgement of receipt of the Letter from the COAS Secretariat.

Few months later, GOC of the Division came on a routine visit to our Brigade. He was a thoroughbred professional soldier. I showed him the copy of the Demi-Official Letter just for his information. He read the Letter and looked into my eyes intensely. Very affectionately, he said, "R V, I would not have been able to do this", and the matter ended then and there. No questions were ever asked by him on the subject matter. GOC rose to become a Lt Gen and retired as Commandant, Infantry School, Mhow. Yes, both the accused Maj Gens were removed from Service.

Our system is quite good and takes due care of most of the aberrations as and when they occur. Also, I am of the firm opinion that if one's intentions are good and working for overall good of the Organisation, blessings of God Almighty will always be with you. Yes, some unfortunate cases will always be there and same need to be understood and taken in the right spirit.

# A COCKROACH IN THE PLATE OF MY CORPS COMMANDER

I was commanding a Brigade at Kanpur. Visit by Senior Commanders to lower Formations / Units are a routine and an essential part of Army life. Our Division and Corps HQ both were located at far off places. Hence, interaction with senior Commanders was not a daily routine. Whenever a senior Commander visited the Formation, it was always treated as an important event, hence, planned and executed with due military ethos. Division Commander, located at Allahabad had visited our Brigade thrice in a year and Corps Commander, at Mathura, just once in my entire tenure of two years.

It happened sometime in the year late 2000. Lt Gen R S Kadyan was our Corps Commander. I had already spent six months in command of the Brigade and was well settled. Lt Gen R S Kadyan, Corps Commander decided to visit our Brigade. It was a short visit of a day only and his lady wife was not accompanying him. He was scheduled to arrive in the morning by a helicopter, carry out his scheduled visit to the Brigade, have lunch and fly back to his HQ in the afternoon itself. Division Commander was not accompanying him as he had visited us just a fortnight ago. As a matter of principle, I do not drink liquor nor have I ever offered any liquor to any person at my residence. Also, I always treated guests at Flag Staff House as my personal guests. However, I always offered drinks to the Visitors / Guests as and when they were entertained in Officers' Mess. This had been respectfully conveyed to the Higher HQ and was well appreciated. Since Lady was not accompanying him and his return to Corps HQ was scheduled the same day, lunch for the Corps Commander was organised in Brigade Guest Room where he was to halt for a little while only. Corps Commander and self were the only two attendees for the Lunch.

While visiting the Battalions, Corps Commander decided to travel by a Jeep in place of Staff Car. This caused bit of problem. Since I did not drive Military vehicle, I had to then sit in another vehicle and lead. Corps Commander resolved the issue by deciding to drive the vehicle himself with me sitting on the Co-driver seat. He observed everything very intently, asked very pointed questions and simultaneously cracked numerous jokes to lighten the atmosphere. Visit went off well and it was now time to go to Guest Room for Lunch. We both sat at a Four Seater Dining Table and were waiting for the food to be served. There was a time lag of 2 – 3 minutes before the arrival of the food. Suddenly, I noticed a big Cockroach coming out from nowhere and in no time it was in the centre of the empty Food Plate of the Corps Commander. We both noticed the Cockroach almost together, looked into each other's eyes and again at the Cockroach. By then, Cockroach had taken a full chivalrous round of the Food Plate and now descended on his next destination i.e. big spoon. I maintained a studied silence as nothing much could be done by me at that moment. The uninvited guest vanished in thin air as it had emerged a little while ago. Entry of Mess Havildar and a Waiter, laden with food to be served, brought much-needed relief. I told Mess Havildar to change the entire crockery and cutlery with fresh one before serving the food. Corps Commander had a hearty laugh once Mess Staff had left for bringing the fresh set of crockery and cutlery. He did not discuss this incident anytime thereafter. We enjoyed the food and he happily returned to his HQ in the afternoon.

It was so very kind of him to behave in such understanding and gentlemanly manner. My respect for him grew many-fold thereafter. May God bless him with a happy, healthy and cheerful life always.

## IMPORTANCE OF HUMILITY IN LIFE

Our ancestors, epics, and all religious texts have always emphasized the need to be humble in life. It becomes further more desirable when one reaches a position of authority where your unbiased actions and instructions can make a sea of difference in the well being of people around you. Humility should always be cherished and, on the other hand, ego should have no place in a balance and mature human being. If one does not respect this maxim, sooner or later, or may be in some future life, surely pays the equivalent price of our negligent behavior.

I took over an Infantry Brigade in Kanpur in May 2000. Higher HQ had formalised a little improper system for local administration of the troops located in Kanpur Military Station. Major troops in the Cantonment were from an Infantry Brigade, an Armoured Regiment and a Central Ordnance Depot. Somehow, it had been made as a policy by higher HQ that Commandant of Central Ordnance Depot will always be the Station Commander, thus leaving no leeway in the hands of Infantry Brigade Commander to execute matters for local administration of his troops. This system had number of shortcomings at ground level and created avoidable acrimony between the two Commanders. Often, there was a running feud between the two Brigadiers on the subject of allotment of funds for execution of Works by MES to Brigade Units vis-à-vis COD/ Station Units. There were few other Station matters, like, management of Kanpur Club which also contributed in aggravating the matters. This state of mistrust had been festering for quite some time. On taking over command of the Brigade, I decided to start with a clean slate and not to be guided by past happenings. With this thought in mind, I planned to interact with Commandant, COD who was also Station Commander since last over two years.

It was Sunday morning, around 10 am or so, when I phoned him up. After exchanging normal pleasantries and self introduction, I respectfully conveyed my desire to have good relations with him on all matters, be that official as well as personal. Brigadier V S Yadav, Commandant COD listened to me and his reply left me astounded. He said, "Look Brigadier, please, remember that the hand of a GIVER always remains at a higher level than that of a TAKER". I was left speechless by the audacity of the person and his inflated ego. Anyway, I controlled myself and let the conversation end on a happy note. After the conclusion of the call, I tried to analyse his personality and understood him to be a highly egoistic person. I had no option but to manage things with him. Alas! Fate had other things lined up for him and so soon.

Around 12 am the same day, I got a call from Major General Administration at the HQ Central Command, Lucknow. I was surprised to receive this call from him on a Sunday. Also, it was against normal protocol of service to receive a call directly from Command HQ unless it was urgent. He informed me there had been a major outbreak of fire in COD Kanpur two hours ago engulfing storage sheds which contained costly and strategically important items of High Altitude Warfare, meant of troops deployed in SIACHEN GLACIAR. As per him, majority of the stores stocked in the sheds appeared to have been burnt in the fire. He conveyed to me that I have been appointed as Presiding Officer of the Court of Inquiry and asked to proceed to the site of incident immediately.

Initial inspection of the site revealed negligence in storage of items, holding of Fire Fighting Equipment and regular conduct of Fire Fighting Practices. Losses occurred were large and of strategic importance and recouping would take quite some time. It was a case of overall negligence by the holding Depot as well as senior technical supervisory Staff at Command and Army HQ levels. I spoke to MG Administration and requested him to upgrade the level of conduct of Inquiry to a Major General as overall negligence and indirect involvement of Commandant COD was also suspected. This was agreed upon; Commandant was removed from all responsibilities and attached with my Brigade HQ. Soon a new Commandant was posted in his place. What an irony of fate and so fast!

Court found almost all dealing senior appointments of COD Kanpur were involved in the negligence and various irregular activities and blamed for their lapses. Some of the persons blamed deserted from service while others were duly punished.

Now, a few words about the fate of the old and new Commandants of COD Kanpur. With in a very short time of occurrence of this fire incident, old Commandant was detected having Cancer and died soon after. Second Commandant became a Major General, was court-martialed for certain irregularities and fled to a neighbouring country before promulgation of the punishment. God knows what happened to him thereafter

In my personal opinion, it is better to be humble and polite and still be correct and righteous all the time. Being egoist never pays, so why carry this burden with oneself; instead shed it at the earliest. Good will and unseen blessings of all colleagues – seniors and juniors alike – will always help in the long run. God Almighty, in any case, is always there to help us.

# LITERACY VS EDUCATION - WITH MALICE TOWARDS NONE

'Literacy', the ability to read and write, or AKSHAR GYAN in Hindi, is often equated with 'education' but in my humble opinion, it is not the same. For me, education is the complete development of a person, or shall we say his personality, in terms of knowledge, sensibility and his behavior under varied situations. A literate individual has to internalize what he has learnt as part of his 'literacy' campaign. Please remember, 'literacy' can be forgotten but, once internalized, not 'education'. I happened to undergo not so a savory situation in my younger days as a Captain and learnt the difference between 'literacy' and 'education' under rather unpleasant circumstances. Let us reminisce how the situation unfolded itself.

Battalion moved from Mizoram to Trivandrum in June / July 1976 - from Eastern most part of the country to its Southern-most tip. I had bought a new VESPA Scooter through CSD Canteen, allotted to me after a long wait of six years; those were the days when one had to wait for six years to buy even a Scooter. Battalion got busy initially settling down. A few months passed before I realized I needed to have a Driving License to take my proud possession beyond the precincts of Pangode Military Cantonment at Trivandrum. Wait for Driving License got extended as we all proceeded on few months long Desert Orientation Training to Rajasthan and returned in October 1976. It could not be delayed any longer and I now applied to the local RTO for grant of driving license. I managed to pass the requisite tests and was asked to come after a fortnight to collect the license.

It was sometime in end November that I visited the office of RTO for collection of my license. As expected I was in my Military Uniform, a young Army Captain from a Gorkha Battalion vying to look at his best, went straight to the office of concerning Clerk and requested him to give my Driving License. Clerk appeared to be busy in his files as also there were few other persons around him. After few more minutes of wait I reminded him. My reminding appeared to have irked him. He lifted his attention from the Files he was busy with, looked at me menacingly and shouted," BHAGO BHAGO, MAIN HINDI NAHI JANTA". Now, a piquant situation arose. I, an Army Captain, in full Military Uniform, speaking to the Clerk of RTO Office in English language and he in turn shouting at me in Hindi saying I should go away as he does not understand Hindi language. There were RTO Office Staff as well as other visitors around at that time. I told him why he is shouting. Apparently, from my way of speaking English language, the clerk had realized I am not from Kerala but from some Northern State, thus reacted in an ungentlemanly manner. I did not want to escalate the matter, instead went to the Regional Transport Officer to apprise him of the misbehavior by his lower staff. He was busy having endless cups of tea with a group of people. Every time I tried to enter his Room, I was asked to wait. At last, after almost half an hour of waiting outside his office, he gave me a short hearing. He did not even call the concerning Clerk to his office in my presence and simply told me that he will look into the matter and I may now leave. I felt insulted by the perfunctory manner in which the RTO disposed off my complaint of outright misbehavior by his subordinate staff. I had no option but to leave his office and returned to the Battalion very perturbed.

I thought over what I shall I do to redeem my honour. I did not discuss the matter with anyone in my Battalion. After few days, I wrote a letter to Mr Achyut Menon who was then Chief Minister of Kerala. First, I reminded him that, on his request, two contingents from our Battalion had participated in Independence Day March Past that year whose salute was taken by him and I happened to lead one of those two Contingents. Then I narrated complete details of the case and asked for investigation and appropriate punishment to the defaulter. In the end I lamented that though literacy rate in the State is quite high yet its residents neither appear to be properly educated nor understand the true values of living a respectful life in the society. Chief Minister ordered an Enquiry headed by an Inspector General of Police and the defaulter was immediately suspended pending conclusion of enquiry. I was called for giving evidence in the IG Office. Civil Administration was routing the correspondence of asking me to attend the Court of Inquiry proceedings through local Station HQ. Administration Commandant tried to scare me how dare I wrote a letter to the Chief Minister directly bypassing laid down channel of command which will have serious repercussions. I replied nothing to him and simply collected the letter asking me to visit IG Office for giving my statement. Leaders from State Government Employees Union visited my house twice requesting me to withdraw the application which I did not. After some time I got posted out to a Brigade HQ at Dharamsala ( HP ), and never really kept track of the case.

I always felt that I did the right thing by writing to the Chief Minister of the State and brought the incidence to his notice for a remedial action. It was so kind of the Grand Old Man - one of the three main architects of Leftist ideology in Kerala ie Mr EMS Namboodaripad, Mr Gopalan and Mr Achyut Menon himself, to rise to the occasion. His esteem in the eyes of a Young Army Captain went high and I cherish his prompt action on the matter with regards. Yes, I still maintain that AKSHAR GYAN / HIGHER LITERACY is no replacement of TRUE EDUCATION which enriches the society with basic values of life. I have written this WITH MALICE TOWARDS NONE and hold a large number of my friends from Kerala in high esteem.

# SOME OLD MEMORIES - 2/LIEUTENANT RAVIN KHOSLA

It happened long back, sometime in 1985-86. Battalion was located at an ever-clouded yet beautiful place, called BAISAKHI in KAMENG Sector of Western Arunachal Pradesh. We had a number of happy-go-lucky, yet wonderful officers serving with the Battalion those days. A very nice team. Few of them had more than one generation serving with the Battalion. Some out of them later commanded the Battalion and even rose to become Brigadiers and Generals too. I am proud of my long association with all of them.

One day, a young officer joined the Battalion on commissioning from Indian Military Academy, Dehra Dun. He belonged to Delhi and got his education from Doon School. Short in height, like many other officers who generally opt to join Gorkha Regiments, thinly built and had intense looking eyes. We became my room-mates; one may term it as a room though it was nowhere near one. He had no previous Army connection. My new roommate was not fond of drinks, not at least those days what I saw of him. Those days, it was common to have endless drinking sessions in the Officers' Mess till late night. I noticed a sure discomfiture in him of those regular drinking sessions which he could do nothing to avoid. I thought over as how to help this young officer to some extent.

After a few days, we two had a short informal discussion. He was soon to proceed on Young Officers Course, conducted at Infantry School, Belgaum and should prepare himself accordingly. Battalion had a big electric Generator to provide electricity to the Battalion which was generally switched off at 10 pm. I told him to buy a Petromax for his night studies as also collect some GS Publications from Adjutant which he promptly did. I had been Instructor in Young Officers' Wing at Infantry School, Belgaum and gave him a small briefing and few booklets which I still had. He soon organised himself and appeared to be bit motivated as I often noticed him reading something or other.

I soon moved to local Brigade HQ for 2-3 months to officiate as BM as the permanent incumbent had proceeded on a foreign course and moved out without relief. Soon after, I got posted as DQ of an Infantry Brigade at Babina. Subsequently, on promotion as a Lt Col, I moved as GSO1 (Operations) at a Division HQ and then to command another Battalion of the Regiment. Hence, my association with that Young Officer in the Battalion was rather short. We did manage to keep track of each other all along.

We both progressed in our life career and it was 20 years since I met him last at BAISAKHI in Arunachal Pradesh in 1985-86. I had retired as a Brigadier in 2003 and moved to Bhopal as Director, Ex-Servicemen Welfare with Madhya Pradesh Government. Young Officer, by then, had become a Colonel and commanding another Battalion of the Regiment at a far off place, Port Blair, in Andamans. He invited us to visit him in the Battalion he was commanding which we did in the year 2005. We enjoyed his enormous simplicity and hospitality in the Battalion Officers' Mess as well as at his home. He organised a Dinner in our honour in Battalion Officers' Mess and I still vividly remember how he introduced us to his Battalion Officers and Ladies that evening, " Ladies and Gentlemen, Brigadier R V Singh is the same person I have been often talking about and who sent me a Petromax, other study material and motivated me to organise my studies to prepare myself for Young Officers' Course when I was a 2/Lieut". Not very many people say that; more so for a retired officer.

I am proud of my association with that 2/Lieut who is now a serving Lt Gen and our present Colonel of the Regiment. May God bless you, dear LT GEN RAVIN KHOSLA. May God be with you always.

# SAHEBJI, SAB IKKO HI TOH HAIGA

It was sometime in April 2000. I was serving in Corps HQ at BB Cantt at Srinagar and had received posting order to take over command of a Brigade at Kanpur. I had a MARUTI 800 Car which I had myself driven to Srinagar while moving on posting two years ago and had a nasty experience of driving through Banihal Tunnel. There was lot of slush and numerous stones in the Tunnel and my Car almost got hung up on few stones. Numerous vehicles behind blowing horns incessantly almost unnerved me. Some divine help came to my rescue and Maruti Car decided to move down from the stones and we safely drove out of the Tunnel. However, the incident counseled me to take help of some experienced driver in driving down from Srinagar to Jammu whenever I move out on posting from there.

On receipt of my posting to takeover a Brigade at Kanpur in April 2000, I was reminded of my previous experience at Banihal Tunnel. I requested a friend who was commanding an ASC Battalion at Srinagar, to provide me an Army Driver. He was to drive my Maruti Car from Srinagar to Jammu where from I was to take it on. I requested my friend to send the Driver to my Office for necessary briefing. Naik Jagdish Ram Sharma has been detailed and was scheduled to come to my office at 11.00 AM next day for necessary briefing. I instructed Office Runner to be on the lookout for this person and bring him to me when he arrives. I waited till 12.30 PM next day and there was no sign of Naik Jagdish Ram Sharma anywhere. I again spoke to my friend who confirmed Driver indeed has been sent. Office Runner again looked around and did not find the person. However, he intimated that, instead of JAGDISH RAM SHARMA, a Sikh Soldier from ASC had been standing nearby since quite some time. I asked him to send the Soldier to me. I enquired if he had been sent by my friend for briefing. He confirmed he was Naik Jagdish Ram Sharma and sent for briefing. I intently heard his name, looked at the smart Turban which he had proudly worn on his head; it left me little perplexed. A tall, handsome, turbaned Khalsa and bearing name as JAGDISH RAM SHARMA. I thought of going into further details. I respectfully asked him few queries which he readily answered.

JAGDISH RAM SHARMA belonged to a village in Patiala district and had a prosperous agriculturist family behind. During 1980s, Punjab was in turmoil of disturbances caused by the KHALISTANI MILITANTS against the Hindus all over the State. Patiala was no exception and a threat had been issued to the family of Naik Jagdish Ram Sharma too to either convert to Sikhism or sell their agricultural land and house to other persons of the village and migrate outside Punjab. Non-acceptance of the terms would face sure elimination at the hands of the KHALISTANI MILITANTS. After lot of deliberations amongst the elders, entire family chose to convert to Sikhism. My next obvious query was whether any thought of reverting back to Hinduism ever came to their mind since normalcy had returned to Punjab long back. JAGDISH did not wait even for a fraction of a second in giving his reply, "SAHEBJI, SAB IKKO HEE TOH HAIGA, KOI FARQ NEE HAIGA". His reply did not surprise me at all. It made very proud of being an INDIAN and born and brought up in INDIAN CULTURE. May WAHE GURU bless you my dear NAIK JAGDISH RAM SHARMA, A PROUD SIKH WARRIOR, where ever you are.

# PART – III

# SOME PARANORMAL EXPERIENCES

# AN ENCOUNTER WITH A LADY GHOST IN HOTEL SAVOY, MUSSOORIE

HOTEL SAVOY in Queen of Hills, Mussoorie has been one of the oldest and well-known destinations of high and mighty from all walks of life during the British era and post-independence days. It may even surprise the uninitiated that the Hotel has residential blocks named after prominent personalities of yesteryear, like, Moti Lal Nehru, G D Birla and an industrialist named Narang since they were the principle patrons during golden days of the Hotel in the previous century. Savoy has been the most sought after resting nest of many prominent Hindi film personalities too, since olden days. However, Hotel went in decay, ultimately closed down in mid-1990s and remained so for almost two decades. It was during this closure period that the stories of Hotel being haunted by Ghosts went viral all around. This eminent property has been made operational since 2013.

Stories on Hotel being 'haunted' existed since British era. This had been based on an incident in which a British couple stayed in the Hotel in mid 1930s for a few days during which the lady died under suspicious circumstances. Investigations revealed the lady died of poisoning yet no one could be blamed for lack of any concrete evidence. Ever since this incident, it was frequently reported by a number of guests and the permanent staff, of having seen a lady in white robes walking in the corridors of various residential Blocks of the Hotel at midnight. No harm was ever done by this figurine to any Guests or Staff of the Hotel. Sightings of the Lady Ghost in white robes were numerous and frequent and it was but natural that a large number of people, in and around Mussoorie, believed the occurrences.

Hotel changed hands in mid 2000s when it was sold to a Kanpur based Business House who happened to know me. I moved to Hotel Savoy in Mussoorie in Sept 2011 to assist in carryout essential liaison and coordination with various agencies of the State Govt as also be a representative of Owners of the Property at the Site. I occupied a so called 'Suite' in Nehru Block, located on the Western extremity of the Hotel. Entire property was in rather dilapidated state, roofs caved in, bathrooms fallen down, retaining walls collapsed, wooden floors worn out and a horde of stray dogs almost owned the property by night and proclaim so by continuously howling and hooting. Also, the property had become a vast playground for numerous bands of Monkeys and Langoors by day which Mussoorie is so very famous for. Decayed roofs were their most sought after place and kept jumping from one block to other at will. It was a common sight of mother monkeys jumping from one residential Block to other so confidently with baby monkeys clinging on to them. It was quite common to experience a pair of monkeys climb on the roof of Nehru Block at mid of the night, jump around the broken roof and create a ruckus at such an odd hour. It was here that I got to see how an old pair of mother monkeys took control of baby monkeys, bring them to a suitable tree, teach them how to climb, jump from one branch to other and, if required, give a tight slap to a hesitant young one. A newcomer to the Hotel those days always got a scary welcome from the monkeys and Langoors by day and numerous unfriendly canine friends howling and hooting during the night. Since most of the floorings all over the property was made of wood and broken, it became a convenient resting place for the dogs during day as well as night. One often experience a bit creepy feeling while climbing from Library Chowk to Nehru Block after sunset and passed through the main Hotel complex. It came but natural to step faster so as to reach Nehru Block earliest. Hotel had hired 30 odd Nepali Labour during the renovation stage of the Hotel and all were living in the temporary Tin Sheds constructed within the Hotel complex. Some unknown person giving a kick at the backside of an individual found relieving himself in the open in the Hotel area at night was often a hot subject of talk next day morning. One day it became difficult for me to convince a young man from some far off place that the Hotel is not haunted by the Ghosts. He had heard from someone there is a Hotel in Mussoorie which is haunted and he specifically came to Savoy to confirm that. It took me some persuasion to convince him Hotel was not haunted. Such incidents caused further interest in my mind to know what the truth was all about the Lady in the White Robes seen wandering in various corridors of the Hotel at mid of the night.

Nehru Block had two Floors, each having seven Suites and a wooden stairs leading to first floor at each end of the Verandah. My 'Suite' was on the right extreme of the First Floor, closest to the wooden stairs. Two of my colleagues from Administrative Staff were staying in the adjacent room whereas two rooms at the farthest end of the First Floor, were reserved for the Owners. Each suite was quite big in size and catered for the requirement of a family. Motilal Nehru used to book this entire Block for his entourage of family and friends for their prolonged stay at Mussoorie. Hence, this block was called NEHRU BLOCK by the Hotel Staff.

At the very beginning of my stay with Hotel Savoy, a local friend presented a statuette of Goddess DURGA which had a string of small LED lights twinkling. I reverentially placed it in one corner of my Bedroom and never switched off its twinkling lights even by day. A lone person does need an anchor, more so when residing in a rather dilapidated property inhabited by monkeys, Langoors and numerous bands of stray dogs. Goddess Durga was a perfect sheet anchor under all circumstances - good or bad, auspicious or inauspicious.

The Suite I was living had three distinct parts. It had a largish drawing-cum-dining room in front, followed by an equally large bed room and at the end were bathroom, toilet and a change room. All doors in the front portion of the Drawing-cum-dining room and bedroom had been fitted with see-through glass panes in the upper half, as was common in the olden days. On the whole the Suite bespoke of the old grandeur of the Hotel. All this past glory, lying in such a truly pathetic state will induce any lone occupant of a Suite to feel lonelier. Yes, as a matter of routine, I always kept tube lights in drawing-cum-dining room in the front and bathrooms in the rear switched on during the night. It was only in my bed room that I used to switch off the lights before going to sleep. Also, there were sufficient number of Tube Lights in the wooden Stair Case and long corridor in front of all Suites on the First Floor to keep the area properly lit up at night.

It was on one such occasion that I too experienced the presence of benign Lady Ghost in the corridors of Nehru Block. In mid December 2012, my two colleagues occupying adjacent Suite had left for Kanpur for a short duration and entire Admin Staff was living on the Ground Floor. I happened to be alone at the entire First Floor of the Nehru Block that fateful night. As usual, I had my early dinner at 9.30 pm and went to sleep.

Mussoorie Hills experienced early snowfall in the winters of 2012. It started snowing in end-November and again in mid-December. Second snowfall was a heavy one and lasted for three continuous days. The incident I am narrating happened during one night of this second snowfall. It was sort of a snow storm and snowfall continued the whole night. Heavy snow was pouring down and winds howling continuously without any break. All doors, windows and loose CGI Sheets on the roof were rattling in the fast blowing winds and making strange mix of creaking and rattling sound and not far away from me. Since I am a sound sleeper, still managed to fall asleep in the midst of all these creepy sounds.

Around midnight, I heard someone knocking at the glass panes of my Drawing Room door which opened in the corridors. Knocking was done delicately, not in a blunt manner, yet was done persistently, rhythmic and without any halt. Also, the knocking was not disjointed and it followed a set pattern, say, THAK-THAK, THAK-THAK, THAK-THAK and THAK-THAK-THAK , THAK–THAK-THAK, and it went on and on continuously. I was still lying in the bed and fully awake by now. I realized this is not the creaking sound of any door or window or rattling sound of any loose CGI sheet fluttering on roof top in the heavy snowstorm blowing outside. It was surely the sound of knocking on the glass panes of my Drawing Room in front of Bed Room by someone standing in the Verandah with an aim to wake me up to open the door. Also, during the entire process I did not observe any pulling or pushing of the front door where knocking was taking place nor further inside at the door of my bedroom. All this activity was happening only at the front door of my Drawing Room which opened in the Verandah. Knocking had been occurring for approximately 15 to 20 minutes. I sat up in by bed and switched on the tube light of the bed room. Strangely, moment the tube light in the bedroom was switched on, knocking at the doors also stopped. I got up from my bed, went to the bathroom, came back and again lied down on my bed. Moment I switched off the tube light, knocking again started and in the same delicate, rhythmic and persistent manner. Now, I was fully sure that the person knocking at the door of my Suite is none other the Benign Lady of the Hotel Savoy who is often sighted in the corridors; tonight she is on her round of Nehru Block and wants to pay a visit to me. As I was now sure who was knocking at my door, the Benign Lady of the Hotel Savoy, there was nothing to worry about. I kept lying in my bed enjoying the persistent, rhythmic sound of knocking till I went to sleep again and got up next morning only.

In the whole experience, I did not encounter anything physically. I continuously heard sound of knocking at the doors of my Suite from the direction of Verandah. Whosoever was the person knocking at the front door of my Suite made no effort to call me or enter inside the room. The person just remained outside only. I got up next morning and entire occurrence of last night was as fresh in my mind. I checked up if there were any footprints in the snow leading to the Nehru Block during the night or in verandah or wooden staircase. There was none. Same was confirmed by the Admin Staff living on the Ground Floor. I checked up if it was sound of some insects hitting at the window panes, in that case there would be few of them, dead or alive, lying near the entry door to my Suite. There were none.

Whomsoever it was, definitely a friendly soul. It caused no harm to me nor made any endeavour to frighten me. I presume it was the same benign soul, wearing white robes, often went around walking in the corridors and often sighted by numerous people in the past. Apparently, she took a round of Nehru Block to interview me personally on that fateful snow storm laden night. I never discussed this incident with any person till I remained in Hotel Savoy as it may have caused rumour mongering by some person, thus damaging the name of great institution named Savoy Hotel. Later, after almost 3- 4 years of its occurrence, I shared it with Shri Ganesh Saili, the home grown writer, a well known and respected personality of Mussoorie and one person who played a stellar role in the renovation and rejuvenation of Hotel Savoy to what it is today. A most prolific writer of repute, he is a store house of information on Savoy Hotel, Mussoorie and Garhwal Hills.

An interesting, real life experience to go through at mid of a snow storm laden night at an old, dilapidated and supposedly haunted place. Memory is as fresh as if it had happened just yesterday. May God grant SADGATI to the wandering white robe soul of Savoy Hotel.

# VISIT BY A GHOST FAMILY TO MY ROOM IN A HOTEL AT RATLAM

After retirement from Army in the year 2003, I moved as Director Ex-Servicemen Welfare, Madhya Pradesh and Chhattisgarh for a period of four years. Nature of work took me to each every District of the States at least once in a year. Accomodation for my stay was generally arranged in the Circuit House or a Hotel, preferably the former, at the District HQ Place. Sometime in early 2005, I was proceeding on my annual visit to Mandsaur and Neemuch in Western Madhya Pradesh and required to spend a night en route at Ratlam. Since no accommodation was available in Circuit House, arrangements for my stay had been made in a Hotel.

It had been a long haul from Bhopal to Ratlam all the way. I decided to have early dinner and go to sleep. I have a strange habit; whenever I am away from home and alone in a room, always keep a light switched on in the bedroom or bath room before going to sleep at night. This is mainly to avoid hitting some furniture items at a new place while going to the toilet at night. Here also, I had kept a bathroom light switched on and bathroom door left ajar. My bedroom had two single beds separated by small bedside table in the middle on which were kept two glasses and a water jug. My bed was adjacent to the wall of attached bathroom. Other side of the second bed was a small round table and two easy chairs. Behind these chairs was a window opening outside. I am an 'early to bed and early to rise' type of person and go to sleep around 9.30 or latest by 10 pm regularly. Also, I am a sound sleeper, go to sleep with in 10 seconds of closing my eyes and generally do not wake up unless required to use bathroom.

Around midnight, I heard some people talking and they appeared to be inside my bedroom. Though I was still not fully awake yet could make out there was one male and a female voice talking to each other. Later, I heard voice of a young child too who may be just 7 – 8 years old. I presume they were husband and wife with their son. Though still half awake, I noticed that the bedroom tube light which had been switched off by me before going to sleep, had been switched on and the bedroom appears to be now well lit up. Half of my face and one eye were covered by the sheet and remainder half of my face and other eye left uncovered. I could well feel the bright light falling over my uncovered eye while listening to the talk going on amongst the strangers in my bedroom. I kept both my eyes closed and made no movements of my eyes or any body parts whatsoever.

I was neither fully awake nor fully at sleep. It was something between the two. I was definitely not dreaming. I heard the voice of two adult people talking in my bedroom. It appeared both the adults were sitting on those two easy chairs and talking. Sometimes I could hear the child also interjecting in their talk. Twice or thrice I heard the male voice telling the female that the person, that is me, is awake; every time the female counter replied in negative saying the person was fast asleep. Male addressed the female by a Muslim name. I have forgotten her name now though that remained etched in my memory for many years as I prayed for grant of peace to these departed souls who had intruded in my bedroom that night. After sometime, female went to the bathroom and I clearly heard the sound of Flush Tank being used. Little later male went to the Bathroom and I again heard the sound of Flush Tank being used. I experienced their stay in the bedroom for 15 to 20 minutes or so. Suddenly, I could hear nothing and there was a total silence thereafter. Apparently, the entire family had left; I fell to sleep again.

I was not fully awake in the initial stage of this occurrence nor was I in full sleeping state. As said earlier too, I was definitely not dreaming. I was almost awake during the latter half of the occurrence. I did not feel scared of presence of some unknown persons in my room and nor of their sudden disappearance from the scene. They neither disturbed me in any way nor caused any harm to me. Even the entire setting of bedroom furniture and other items was left undisturbed. In my humble opinion, they were some benevolent souls who just, per chance, happened to visit my bedroom that night.

I got up at 5 am; incident of last night was still quite fresh and clear in my mind. Generally, one forgets dreams of previous night within a few minutes of waking up next morning. This was not the case with this incident and the whole incident remained very fresh in my memory. Obviously, I did not dream of this incident. I immediately called Duty Manager of the Hotel to my Room. I asked him if any of the previous occupants of any room of this Hotel in general and this room in particular, has ever reported occurrence of any abnormal activity in the night. His answer was firmly in negative. Then I narrated the incident of last night which I had experienced. He was quite surprised. I left the matter at that and proceeded on my onward journey to MANDSAUR and NEEMUCH little later.

On my return journey, I again halted at Ratlam. This time I was accommodated in Circuit House. I was invited for dinner at a local friend's place and narrated my experience of that night in the Hotel to my hosts. He had a hearty laugh and told me that, during construction stage, one floor of this Hotel had collapsed thus killing the Construction Engineer on the site. Deceased person was Muslim by faith. His wife could not bear the loss of her husband and she too soon died.

I am sure my sincere prayers to the God Almighty for grant of peace to the departed souls had long been accepted and they rest in peace for eternity. All those souls who visited my room in the Hotel that night in Ratlam, did not disturb or caused any harm to me and were benign. A great experience to have gone through and difficult to convince other people of veracity of such like occurrences.

# A CREEPING HAND UNDER MY NECK IN CIRCUIT HOUSE AT REWA

It happened sometime mid 2005. I was on an annual visit to District Sainik Welfare Office, Rewa and arrangements for my stay had been made in Circuit House. Circuit Houses had been made by the British all over India so as to provide appropriate accommodation facility to government officials while on tour to the various Districts. Rewa happened to be an important princely State in Baghelkhand region of Northern part of Central Provinces ( now called Madhya Pradesh ) in pre- independence era. Circuit House at Rewa is an old looking colonial structure with an Annexe of additional rooms abutting the main building to accommodate more visitors when required. I was allotted a room in the Annexe.

It so happened there was no visitor other than me, occupying any room in the entire complex of Circuit House on that day. Room allotted to me had two single beds separated by a small table and an attached bath room. Keeping in mind mosquito menace in rainy season a mosquito net had been duly tucked around my bed. I usually go to bed early, finished my dinner in the Circuit House around 9 pm, retired to bed by 9.30 pm and was soon fast sleep. I left the tube light of attached bathroom switched on for the night as is my habit while I am away from my home.

Around mid-night or so, I felt a hand creeping under my neck and trying to lift my neck up to make me sit up. I was sleepy and did not sit up and went to sleep again. While still half asleep, I again felt the creeping of a hand under my neck and trying to raise my neck and back to make me sit up. This time more force appeared to have been applied by the person whomsoever it was. I sat up in my bed and was rather surprised on the occurrence. I intently looked around in the room for the person who was trying to make me sit up. I found none other than me in the room. Also, I found mosquito net had not been disturbed and was as well tucked all around as was prior to my going to sleep. I came out from the mosquito net and checked the bathroom. Latch on the bedroom door was in closed position and had not been touched by any one. It was mid night, I came out of the room and walked around the entire Annexe Complex of the Circuit House for a while. There was total silence all around and found no one awake other than me. All members of administrative staff of the Circuit House were also fast asleep. I returned to my room and went to sleep again. I did not experience any further such like occurrence in my Room thereafter during the remainder part of the night. I slept peacefully and I woke up next morning only. I did not discuss this incident with any of the administrative staff of Circuit House at Rewa and next day morning moved on to Sidhi, my next scheduled destination of the tour.

I very vividly remember a hand slowly creeping under my neck twice that night as if trying to wake me and sit up. May be someone was trying to interact with me but due to some reasons, could not and left it midway. It caused no harm to me. May God grant SADGATI to the benign Soul.

# PART – IV

# ADVENTURE AND SPIRITUALITY

## SWARGAROHINI (CLIMB TO HEAVEN)

## A TREK TO SATOPANTH TAL IN GREAT HIMALAYAS

SATOPANTH is an aberration of word SATOPATH or SATYA-PATH. SATOPANTH TAL is a small triangular shaped, emerald colour lake of approximate size 400m x 400m. It is located at the base of famous CHAUKHAMBA Mountain in the GANGOTRI Group of Himalayas. This lake has no exit to water drained in from various surrounding Glaciers.

ALAKANANDA River flows between two great mountains features named NARA and NARAYANA. NARA PARVAT is on the East of Alaknanda River extending from SHREE BADRINATH DHAM to GHASTOLI in the North. NARAYANA PARVAT lies on the West of Alakananda River right up to LAKSHMI VANA. SHREE BADRINATH DHAM is located on NARAYANA PARVAT ; adjacent to it flows Alakananda River. Satopanth Trek commences from Badrinath and follows Western bank of Alakananda River up to Lakshmi Vana. There is another track which follows eastern bank of Alakananda River from MANA Village and concludes at VASUDHARA Water Falls approximately 6 km upstream. There are numerous natural and man-made caves at the site of Vasudhara Water Falls where many ascetics live throughout the year and engaged in various forms of Penance and Meditation. Track following western bank goes beyond Lakshmi Vana right up to Satopanth Tal and is famously called as Satopanth Trek. This trek finds a mention in SKANDA PURANA, a famous scripture of Hinduism. Once the Great War of Mahabharat was over and having reigned in peace and prosperity for 36 years, the Pandavas decided to make the final voyage. All five Pandavas and Queen DRAUPADI undertook this arduous journey to Satopanth

Tal to wash away the sins of having killed 100 KAURAVAS, their own brethren. All members of the group, less King Yudhishthar, fell lifeless to the ground one by one. Ultimately, only King Yudhishthar, eldest among the five brothers and a very righteous person who always followed DHARMA / SATYA, survived. Hence, this trek had always been referred as SATYAPATH, got aberrated to SATOPATH and present day called as SATOPANTH. The aim of all Pandavas was to go to Heaven, this Trek is also referred as SWARGAROHINI TREK, a 'Climb to Heaven'.

This trek is approximately 27 km long one way and has four night halts en route, namely, LAKSHMI VANA, SAHASTRADHARA, CHAKRA TEERTHA and SATOPANTH TAL. It may be news to the uninitiated that there are no administrative facilities for boarding / lodging, food, medical support and communication with outside world available beyond Badrinath. Once you step ahead of Badrinath, your faith in God Almighty is your only support to finally bank upon.

Distance between various night halts is as follows :-

| | |
|---|---|
| Badrinath to Lakshmi vana | 8 km |
| Lakshmi Vana to Sahastradhara | 6 km |
| Sahastradhara to Chakra Teerth | 6 km |
| Chakra Teerth to Satopanth Tal | 7 km |
| Total distance (One way) | 27 km |

Badrinath is located at the height of 10,200 ft where as SATOPANTH TAL touches approximately 16,200 ft. Since no administrative support of any kind whatsoever is available beyond BADRINATH, it is more prudent to join a group of trekkers under overall arrangements of an experienced Tour Operator from area around Joshimath. I formed part of a 10 member Trekkers' Group which consisted of nine members from West Bengal. All of them were professionals working in various civil organisations, passionate trekkers and had been regularly undertaking at least one to two treks in Great Himalayas every year. Most of them had already been to famous ROOP KUND, KUARI PASS, GAUMUKH, TAPOVAN and many other high altitude trekking spots in Garhwal, Kumaon and Himachal Pradesh. All members were well cultured and always respectful to me. I was the oldest trekker amongst them, running in my 69 years of age in the year 2016. The majority of them were between 30 to 40 years. Age difference between me and the youngest member was 40 years and every one paid due respect to my age. Some were even motivated to request their parents to come along for their next trek.

We all got together at Badrinath on 23 September 2016 and commenced our pious journey toward Satopanth. Lakshmi Vana was to be our first halt. Our Tour Operator, Manish Bhujwan, had made good arrangements for our administration. He had a band of seven Nepali Coolies to take care of our snow tents, cooking utensils and rations and other stores. I had hired a local person from a village named Tapovan near Joshimath as personal Coolie for carriage of my belongings. Remainder all members of the group carried their personal belongings themselves.

I had undergone 4 days long higher stage acclimatisation schedule at Badrinath to make myself fit for the tough journey ahead. Also, I had to undergo a mandatory Medical Examination to check my suitability for undertaking this Trek. Unfortunately, I was found medically 'UNFIT ' due to high Blood Pressure and advised not to undertake this Trek. I had come all the way from Noida and with so much of burning desire of fulfilling a life time dream of visiting SATOPANTH TAL. I could not restrain myself and decided against the medical advice. I resolved to proceed ahead irrespective of whatsoever may happen. I had procured essential medicines to provide me some respite in High Altitude effects which would later come handy to me.

Around 9 am on 23 September 2016, we a Group consisting of 10 Trekkers, two Tour Operators and seven Nepali Coolies commenced our move from Badrinath to Lakshmi Vana. We crossed Alakananda River at Mana Village itself and took the track which follows the Western bank of the river. As per SKANDA PURANA where we first find a mention of SWARGAROHINI TREK, Queen Draupadi could not proceed beyond Mana Village and died . Hereafter, only five Pandavas continued with the Trek ahead. After walking for approximately 3 km we found there was no track ahead. Track had been washed away by the glaciers rolling down from the adjoining mountains. It had been further worsened by the just concluded rainy season and many landslides had washed away the remnants of the track alignment wherever it had earlier existed. We waded through 3 – 4 glaciers each width of 300m to 400m and numerous landslides en route. Hardly any alignment of earlier existing track was left and movement ahead was becoming quite difficult. Last 4 km of our route had been through numerous landslides and glaciers only. Mid way, we came across a young couple from Mumbai who were on their way back from visit to Satopanth. They had hired two Coolies for provision of administrative support to them.

We could now see Vasudhara Water Falls on other side of Alakananda River. The sight of water falling from the height of an altitude of 14,000 ft for a distance of approx 1000 ft down below was beautiful. By the time falling water touched the rocky surfaces down it was all vapours and not even a droplet of water in it. We could notice few caves other side where many Sadhus had been living to carry out Penance. Vasudhara Water Falls is regularly visited by the Trekkers and pilgrims in open season. There are no administrative facilities available at Vasudhara Water Falls and all visitors return to Badrinath the same day.

We reached Lakshmi Vana around 3 pm. It is a somewhat rectangular shaped open space of 150mx200m at an altitude of 12,000ft. Vasudhara Water Falls are right opposite Lakshmi Vana with only River Alakananda River separating the two. Standing at Lashmi Vana, one can very well see two big Glaciers joining at its base. Magar Kharak glacier coming from North and Alkapuri Glacier from North Westerly direction of Choukhamba Mountain. Their meeting point at Lakshmi Vana / Vasudhara Water Falls is considered as the point of origin of Alakananda River. This place is also referred as Alkapuri in the Tourist Circuit. About a km ahead we found a water stream coming from the Westerly direction and contained water flowing down from numerous glaciers emanating from Neelkanth Mountain which were encountered by us during the journey next day. This stream also deposits it water at the junction of Magar Kharak and Alkapuri glaciers mentioned earlier.

Name Laksmi Vana may be a misnomer as there are no Vana / forests around. In actual fact, treeline vanishes much before Badrinath and all mountains in this area are barren. Surprisingly, we found many BHOJA VRIKSHA trees at this place. These trees exist only on the Lakshmi Vana side of Alakananda River. In the olden days, the leaves of BHOJA VRIKSHA were utilised by our Sages / Rishis in writing SMRITIS / PURANAS, our scriptures, in ancient days.

We camped at Lakshmi Vana for the night. Lit a Camp Fire at the Site and had our Dinner at the Camp Fire Site itself. Though I had walked all along, yet was not feeling very comfortable. I discussed this with our Tour Operator who advised me to take the High Altitude medicine before going to sleep and take a decision next day morning.

SEHADEVA, the youngest of the PANDAVAS, could not proceed beyond Lakshmi Vana and supposed to have fallen lifeless at Lakshmi Vana.

Next day, on 24 September, I was feeling better. The view of Vasudhara Water Falls, two glaciers coming and joining up, few BHOJA VRIKSHA close by and grand view of Great Himalayas, barren as well as snow covered, right in front, could motivate and invigorate any mountain loving soul to go to any extent. Also, the medicine taken the previous night had its effect and I confirmed to the Tour Operator of my decision to continue with the Trek ahead. We commenced our move from Lakshmi Vana around 9 am and now headed for Sahastradhara, our next night halt. After moving just a km or so, we came across a Hanging Glacier coming down from the direction of Neelkanth Mountain which so far was not visible. Few more such like Hanging Glaciers were seen coming from the same direction little further ahead. Hereafter, we encountered a very steep climb of about 2000 ft which took us almost three hours to reach on top of the ridge line. At places we found some stumps of old Rhododendron trees. It was truly amazing and mesmerising to see huge glaciers hanging from such height and in so close a vicinity. It was so beautiful and heavenly a spectacle. We were to soon see many more glaciers ahead hanging from Neelkanth Mountain. There was no recognisible track ahead as the same had been washed away by avalanches / landslides during last 6 to 8 months of rain and snow. Upon that there were huge rock faces to be circumvented to find our way ahead. Old stumps of Rhododendron trees came handy in providing support to make our way to an extent. However, our Tour Operator was well experienced, had taken Trekkers to Satopanth a number of times in the past, hence, we always found our way ahead without much difficulty.

We had moved three km further ahead towards Sahastradhara and could notice few more glaciers emanating from Neelkanth Mountain and hanging down. Ultra-white frozen ice with fresh snow overlap was shining so beautifully in Sunlight. Neelkanth Mountain was also now partially visible and numerous glaciers hanging down from it were fairly visible. After covering a distance of about six km from our last night halt at Lakshmi Vana, we reached an open area through which flowed a stream of water. This water stream collected water from the glaciers coming down from Neelkanth and deposited it at junction of two glaciers at Alkapuri, near Laksmi Vana, our last night halt. We continued moving up along the water stream for another km or so and found an open space which was sufficient enough for night camping by a group of 15 – 20 persons. There was an amazing sight on to our West as hundreds of frozen and some running water streams were coming down from Neelkanth Mountain. All these streams flow down in Summer season and remain in frozen state for the remainder part of the year. It was a superb view. It is because of so many of streams, frozen and flowing, coming down from massive Neelkanth Mountain, this place is called SAHASTRADHARA, a thousand streams, which indeed it is. We camped here for the night. Soul of NAKUL, second of the brave PANDAVA Warriors, supposed to have left his mortal body at this very place.

We all had a Camp fire at this place too and had our food at Camp Fire site itself. However, my physical condition had deteriorated from bad to worse. I spoke to our Tour Operator and informed him of my worsened physical state. He was now seriously worried and told me if my condition does not improve by next morning, I may have to drop out from the Trek and return to Badrinath. I did not want to return empty handed after having come that far. I wanted to move ahead and complete the Trek. I took another dose of High Altitude Medicine and went to sleep praying to the God Almighty to make me fit by next day morning. I woke up in the morning and found my condition further worsened. My eyes lids were badly swollen up, eye balls become blood red as if blood is going to ooze out any time and felt so much of headache and body pain. There was no instrument with us to measure blood pressure; I could only count my Pulse Rate manually so as imagine corresponding level of Blood Pressure. Lying in the Snow Tent that morning, I must have counted my Pulse rate 8-10 times, it was always ranging between 110 to 115 and sometimes even reached 120. I gave a serious thought whether to proceed further or turn back. I very well knew if something happens to me, there are no means to evacuate backwards. Perforce, I will be left under some big rock till arrangements for any manual evacuation are made. Final decision to return to Badrinath or proceed ahead had to be made by me alone. I gave a serious thought to it and found that my motivation level to visit Satopanth Tal was over-riding all my poor physical conditions. I very well knew this was once in a life time opportunity to undertake this pious. No such second opportunity will ever come in my life again. I set aside all health related constraints and confirmed to the Tour Operator and all my colleagues of my decision to move ahead, come what may. I took another doze of High Altitude Medicine, left everything to the God Almighty – AB CHHOD DIYA IS JEEVAN KA SAB BHAR TUMHARE HATHON MEIN, HAI JEET TUMHARE HATHON MEIN AUR HAAR TUHARE HATHON MEIN - and moved ahead with the remainder group.

Our next night halt was CHAKRA TEERTH, approximately 6 km ahead. We all left Sahastradhara around 10 am, bit late today, as distance to be covered was little less. In fact march of today was the shortest of all days. Also, certain administrative issues were to be resolved by the Tour Operator. There was a tough straight climb for approximate two km in the beginning itself and with my weak physical condition it appeared endless to me. I looked back at massive Neelkanth Mountain and could now very vividly see the top of the Mountain and all glaciers hanging down which had been encountered by us yesterday. Now, right in front of me in the North, and not far, was PARVATI Mountain laden with heavy snow. During last three days of trekking in those barren and lesser frequented mountains, we came across very few trekkers en route, say, just about 3 – 4 trekkers per day. Little ahead of Sahastradhara, we saw a young girl of 12 – 13 years, from Bangalore, who had accompanied her parents on trek to Satopanth and now returning home. It was so nice seeing parents involving their young children on such adventurous pursuits since their childhood itself. After covering a tough stretch of six km or so, we reached an open flat area of approximate size of 200m x 300m which was called as CHAKRA TEERTH. We could have a 360 degree view of all major mountains around from this place. We still had two hours of day light available and our Tour Operator gave us a detailed briefing from this place on all the major mountains visible; general direction of some of the famous mountains which were yet not visible from this area was explained. CHOUKHAMBA Mountain was right ahead, KUBER on our Right and so were many other Peaks going up to Mana Pass and even beyond and partially visible from this place. Sun was about to set and sunset light had lit the top of all snow clad mountains in golden colour. It was a very beautiful sight to watch. We did not notice any bird or animal beyond Lakshmi Vana. No greenery anywhere around. Marmit was the only creature sometime noticed by us. It makes a burrow in the ground as abode for itself. Nepali Coolies talked of having seen a Snow Leopard sometime in their previous visits to this area. They may be true as this is the right place but where will it feed itself from! There should be some mountain goats or other sort of fauna which so far had not been seen anywhere.

The view of Great Himalayas as seen from CHAKRA TEERTH was exceptionally good. To put it HEAVENLY will be more appropriate. One did not feel like closing one's eyes even for a second lest one would miss this beautiful scene spread by the Nature all around. Clear blue sky, snow clad mountains, clouds hanging over them at few places. It was too good a spectacle to watch and enjoy.

We could occasionally hear cracking sound coming from various directions. It actually sounded like dynamite blasts occurring at far distance. Initially, we thought it was some road construction activity in progress by Border Roads. But it was not so. Our Tour Operator explained the sound as actually emanating from cracking of glaciers around us. Soon we got a life time opportunity of witnessing an avalanche rolling down from a snow clad mountain on to our East, closer to Kuber. First we heard the cracking sound on to our East, our Tour Operator pointed to us the approximate direction, we all focused ourselves in that direction and, my God, we noticed a huge mountain of snow rolling down. There was a ridge line between us and the place where avalanche had occurred, hence, we could witness the occurrence of avalanche for just about half a minute only. We photographed and even videographed the grand spectacle.

There was huge barren ridgeline coming from the direction of Kuber Mountain and going down in the direction of Vasudhara. Alkapuri glacier coming from Choukhamba Mountain as well as Satopanth was flowing in between this ridgeline and Chakra Teerth, where we were standing. This barren rocky ridge line had an almost 90 degree rock fall facing Chakra Teerth. Rock formation, as must have occurred since millions of years of emergence of mighty Himalayas, and as visible to us from Chakra Teerth, gave us the impressions of lookalike of some monolith human or animals and others of very vast dimensions. Yes, one may have to supplant some individual imagination to all these figurines presenting themselves before you.

ARJUNA, the Great PANDAVA Warrior, could not proceed any further and soul left his mortal remains at this place.

We spent the night at Chakra Teerth which was at the height of around 15,000 ft. Place was very windy the whole night. It was our third night out since we left Badrinath. I was feeling better in the morning next day, though quite tired. I thought over of my aim, physical condition and how far we had reached by then. I was so close to SATOPANTH and cannot give up now. I decided to continue. It was 26 September and fourth day of our trek. Fortunately, it was again bright and sunny. Sometime it does rain in higher mountains, even during end of September, thus making movement difficult and even risky. Our Group decided to reach Santopanth Tal and return to the present location Chakra Teerth the same day. Reason advanced was non-availability of a suitable enough camping site for our Group at Satopanth Tal. It meant we had to cover two days of trekking distance in one day.

We started our move little early, at 8 am as lot of distance, to and fro, had to be traversed that day. I was feeling quite fit and energetic since morning itself. Obviously, regular dose of High Altitude medicine was showing its positive effect on me. First we had to traverse half a km long somewhat flat distance and then climb on top of a barren and rather steep ridgeline. It took us one half hour to reach on the top. We could now take a grand and closer view of majestic, beautiful, snow covered Choukhamba Mountain from the top of this ridgeline. Also, from here, we could have a full and better view of the entire 5 km wide expanse of huge glacier lying between Chakra Teerth and Satopanth. Entire glacier expanse was strewn with frozen ice, small and big bolders accumulated since long and moving down in its own slow speed. One thing which was not visible to us was any track ahead through the Glacier toward Satopanth. However, our Tour Operator was quite well versed with the area and we had full faith in him. At number of places we found 3 – 4 small stones placed on top of each other. That was the route marking done by the Trekkers preceding us and we better follow that. Entire route was rather difficult, more often we were on our fours to get across a huge boulder or slippery frozen ice surface and next one was soon to follow. Such difficulty in walking was not restricted to some specific short distance but the entire glaciated area of 5 km width presented itself likewise. It was quite difficult to get across the entire expanse of glaciated region. Not only this, we had to return also and follow the same difficult route all along. At last, we saw a red flag fluttering on top of a ridge line ahead, signaling that the Lake is not far now. Some consolation! But there was a steep climb involved between our position in the Glacier and top of the ridgeline, then only we will be able to have a first look at the pious Satopanth Tal. We all continued climbing up and after some struggling time, all of us reached on top of the ridge line.

What a grand sight it was! Now we could see the Great, Pious Satopanth Tal right in front of us. It was of triangular shape with base at the far end and had beautiful emerald colour water filled in it. It looked so beautiful, its beauty cannot be put down in words and the same can only be seen and experienced by a person himself. Words cannot do justice to its true glory. This lake has no exit yet its water is sweet and ever fresh. Unbelievable but true. I noticed the similar phenomenon with the water of most pious LAKE MANSAROVAR near MOUNT KAILASH in Tibet which I was fortunate to visit the very next year in June 2017. Mansarovar Lake has no exit to its water yet its water remains sweet and fresh as ever. I had brought a bottle of water from both these pious places and the same are reverentially kept at my home ever since. Water is still as sweet and pure as was at their original place, 4 years ago.

We came down to the bank of Satopanth Tal which was hardly any distance. A very small structure, depicting a Temple had been constructed at the bank of the Lake by the residents of Mana village, near Badrinath. There are no Idols of any God or Goddess kept in or atop the structure. Just 30 meters away and on to its right was a small temporary hut made where presently a Bengali SADHU was residing and performing his Penance / TAPASYSA. Earlier a NEPALI SADHU stayed for 3 – 4 years, performed his Penance /

TAPASYA and now left for his spiritual journey ahead. Just 300 meters ahead were two small ponds named CHANDRA TAL and SURYA TAL. Both these small Tals/Ponds were visible from top of the ridge, atop which was the fluttering Red Flag was encountered by us little while ago. Just 1 km to 1.5 km ahead of Satopanth Tal, another important area was clearly visible from atop the Red Flag ridgeline. At the base of snow covered Choukhamba Mountain, was a small flat area which was laden with heavy snow / eternal ice accumulation, part huge rocks and was clearly visible from even far distance. It is believed that this ROCKY AREA is the place where Celestial Vehicle supposed to have landed and took KING YUDHISHTHAR to heaven. Satopanth is the place where Mighty Pandava Warrior BHIMA fell lifeless on ground and his soul left the mortal remains. Hereafter, it was King Yudhishthar alone who walked ahead and was lifted in Celestial vehicle and taken to Heaven.

We all spent almost 2 hours at Satopanth Tal and took bath in the holy lake. Fortunately, it was EKADASI, the eleventh day of Hindu Calendar and considered a most auspicious. It is said that on this day, TRINITY of Hindu religion Pantheon, BRAHMA, VISHNU AND MAHESH, come to Satopanth Tal during the night and take bath. Indeed, we had reached Satopanth Tal on a very auspicious day.

We could regularly hear the sound of glaciers cracking around us as we had numerous high mountain peaks in the near vicinity. Satopanth Tal was at an altitude of approximate 16,200 ft and various other high mountain peaks ranging between 21,000 ft to 23,000 ft. Whole area was strewn with live glaciers.

We all were quite tired, yet as per day's planning we had to fall back to Chakra Teerth the same day. Huge six km wide glacier lying between Satopanth and Chakra Teerth had to be traversed back all the way and in day light itself. Now at the time of our departure from Satopanth Tal, our motivation level was high. We moved faster, route was familiar all along and we made it to our Camp Site at Chakra Teerth by 6 pm. It was just about sunset time and, like yesterday evening, we were able to witness the grand spectacle of snow covered mountain peaks lit up by golden colour hue of sun light. It was so beautiful a sight to witness and remember forever.

I was extremely tired, more so by covering two days long march in one day. Yet I was delighted I made it to Satpanth Tal. I embarked upon a difficult SWARGAROHINI Trek when I was running in my 69th year of age and despite numerous health hazards I was able to complete it. Now, I had to only walk back to Badrinath, we all now decided to complete the three days long return journey from Chakra Teerth to Badrinath in just two days. In that we skipped halting at Sahastradhara and instead came down straight to Lakshmi Vana. It was going down all the way and a familiar route and we made it to Lakshmi Vana in a good time. En route we met 5 – 6 trekkers from Karnataka and Mumbai who were heading for Satopanth Tal. Few of them were young working girls and it was heartening to see them being part of this adventurous pursuit.

It was 28 September and as per changed schedule, it was the last day of our Trekking Expedition. We were to finish this trek on 30 September but have been able to make it two days earlier. View of Vasudhara Waterfalls, Kharak Magar and Alkapuri Glaciers, their confluence at Lakshmi Vana, Bhoja Vriksha trees nearby, all were presenting a most beautiful sight. It was here that I had first noticed a strange phenomenon which initially I had not fully understand and was subsequently clarified by our Tour Operator. At the confluence of these two glaciers and about 8 – 10 ft above the surface level of glacier, it appeared as if the mountain side had been cut by some JCB machine in a perfect straight line. This cutting appeared to have gone for miles and miles all along the mountain wherever there was a glacier adjoining it. I had noticed the same phenomenon again from Chakra Teerth to Satopanth. Infact, this had been caused by melting of glaciers which had otherwise remained frozen since times immemorial. Owing to melting of ice below the surface, the glaciers had been sinking down to a lower level year after year. In my estimate, Magar Kharak, Alkapuri and Satopanth glaciers had sunken down by at least 12 ft to 15 ft, if not more, thus creating a serious environmental hazard in the years to come.

We left Lakshmi Vana at 9 am on 28 September and just rolled down to Badrinath. Entire route was well known and all land slide and avalanche prone area just circumvented a week ago. We all were in joyous mood and made it to Badrinath by 3 pm the same day. I moved further down to Chamoli which is the District HQ of the entire area and had night halt in Badri-Kedar Mandir Samiti Guest House. I reached Haridwar on 29 September and back to my home at NOIDA by the evening of 30 September. I was to have returned on 02 October 2016 but came two days in advance due to compressing of trekking schedule. My face had been fully sun burnt by the effect of Ultra Violet rays during 12 days of stay in the higher reaches of Great Himalayas. My wife took few seconds to recognise if I am the same person, her husband of four decades. Two photographs of mine which she clicked immediately on my return from Badrinath, always remind me of my adventurous days on SWARGAROHINI TREK to Satopanth Tal.

It was a great experience for me. I undertook this Trek in my advanced age and fully well knew it was not a right a right decision to proceed on such arduous Trek, more so when there are no proper medical and evacuation facilities available beyond Badrinath. In case an individual develops High altitude sickness / Pulmonary Odema or some other serious ailment while on trek, one can only pray to God Almighty for the miracle to happen. Few years prior to our Trek, a young Trekking enthusiast from Mumbai undertook this Trek and reached up to Chakra Teerth where he developed High Altitude Sickness. Only cure available for the ailment was to bring him down to lower heights. Trekker was a rich person, had foreseen this eventuality and made arrangements for his helicopter evacuation. He was safely evacuated direct to Dehra Dun. But what happens to a common man who is unable to finance a helicopter evacuation? Pray to God Almighty to shower His grace; and He did shower His grace on all of us and during the whole duration of the Trek.

I am grateful to God Almighty who gave me motivation, strength and good health to complete this Trek even during my advanced age. I am ever grateful to my so very friendly and ever respectful nine Co-Trekkers from West Bengal in whose cheerful company I was able to complete this arduous trek. I am grateful to Manish Bhujwan, our Tour Operator from Joshimath, for helping me throughout. It was so unfortunate that he died of Cardiac Arrest in Joshimath within three months of organising this Trek. We lost a good friend in him. May God grant SADGATI to the departed soul.

*Vasudhara Waterfalls Clad in a Rainbow.*

*Lakshmi Vana - Point of Origin of Alakananda River.*

*Satopanth Glacier.*

*Satopanth Glacier.*

*Satopanth Tal.*

*Satopanth Tal.*

*Sight of an Avalanche Rolling Down.*

*Sunset Golden Hue on Satopanth Mountain.*

*Our Trekkers Group.*

*Our Trekkers Group.*

*Two Roy Brothers in our Trekking Expedition.*

*Trekkers and Support Staff.*

# A PILGRIMAGE TO PIOUS KAILASH AND MANSAROVAR LAKE

KAILASH!! MANSAROVAR!! These two very words fill the heart of most Hindus, Jains and Buddhists with reverence. KAILASH and MANSAROVAR are considered as most pious places by the followers of these three religions and a pilgrimage to them a must and ultimate desire in life. Not very many people can undertake this pilgrimage due to physical difficulties and, to an extent, financial constraints involved. We had been taught through our ancient texts since times immemorial that KAILASH PARVAT is the abode of SHIVA, one of the Trinity of Hindu pantheon, and his consort PARVATI. It is the ultimate wish of a Hindu to undertake a pilgrimage to KAILASH – MANSAROVAR once in his life time, pay obeisance at the feet of SHIVA and bathe in Holy Waters of Pious MANSAROVAR LAKE, not far from KAILASH.

We had been contemplating to undertake this pilgrimage since retirement from Army in 2003. I suppose my motivation had not reached the requisite level and the plans kept getting delayed year after year. Meanwhile, in 2016, I undertook arduous SWARGA-ROHINI TREK, literally meaning TREK TO HEAVEN which lies deep in the Himalayas, much beyond BADRINATH DHAM. Undertaking this trek enhanced my motivation level and finally decided to proceed on holy pilgrimage in June 2017. Maximum age limit for pilgrims was 65 years. Death of numerous elderly pilgrims on this pilgrimage had been quite common since ancient times. However, number of cases had increased in the recent past, thus forcing Chinese Government to be extra cautious on grant Visa to the elderly pilgrims. I was already running in 70th year and my wife in her 66th year. Despite being overage, we still went ahead to make a fair attempt and applied for the Visa. We were fortunate as our documents got cleared from the Chinese Embassy for a fortnight long pilgrimage in first half of June 2017. It was matter of rejoice and our preparations for equipping ourselves as also improving our physical fitness started in the right earnest.

In the year 2017, four routes were available to the pilgrims to visit Kailash – Mansarovar and the same are briefly discussed as under: -

1. PITHORAGARH - on foot / by road – LIPULEKH – TAKLAKOT – by road – DARCHEN (Base Station for KAILASH) and back. Duration – 12 to 14 days including 6 to 8 days of difficult climb. Organised under the aegis of Central Govt.

2. KATHAMANDU / NEPALGANJ – by fixed wing aircraft - SIMIKOT – by Helicopter - HILSA – by road to TAKLAKOT - by road - DARCHEN and back. Duration 10 to 12 days. Organised by Private Tour Operators ex Delhi and Kathmandu.

3. Kathmandu – by road – KODARI– DARCHEN and back. Duration – 12 to 14 days. Organised by Private Tour Operators ex Delhi and Kathmandu. Another option ex Kathmandu was to fly from Kathmandu to Lhasa and then go by road or Helicopter to DARCHEN and back.

4. DELHI – BAGDOGRA - GANGTOK – NATHULA – KAILASH and back. Beyond BAGDOGRA, all by road. Duration – 25 to 30 days. Organised under the aegis of Central Govt.

We were a group of 10 PILGRIMS and opted for Option Number 2. It implied us to gather at NEPALGANJ and then take a fixed wing Aircraft to SIMIKOT. Helicopter Flight from SIMIKOT was to further take us to a place called HILSA on NEPAL - TIBET Border and then travel by road to TAKLAKOT, MANSAROVAR, DARCHEN and further on to the base of KAILASH. Visa documents of all 10 of us had been clubbed together and all our movements from NEPALGANJ onwards had to be undertaken as a Group only. Mr V S Kotwal ,76 years and his wife Mrs Jaishri Kotwal, 74 years belonged to Jammu and were the eldest among us. It was followed by me and my wife Kamlesh. Remainder six pilgrims in our group were young people ranging between the age group of 28 to 38 years. Among them were Dr Anupam Goyal from Chandigarh, Mukesh Somani and his wife Anju Somani from Chittorgarh, Basant Jha from Ajmer, Ram Babu from Bharatpur and our youngest member was Nimit Tiwari from Gurgaon.

We all gathered at Nepalganj on 01 June 2017 and were briefed by our Tour Operator about the entire journey ahead, do's and don'ts to be followed and issue of certain items of high altitude clothing to each individual. We were repeatedly instructed not to carry any photo or video clip of Dalai Lama with us and any violation may cause avoidable repercussions. Next day, we all gathered at Nepalganj Airport and waited for the whole day to take an airlift to SIMIKOT. Weather remained bad almost the whole day. Since weather was not likely to improve for next 2 – 3 days, all of us were taken to a Resort at a place called Bardia, 70 km due West of Nepalganj. We remained there for two days and returned on 04 June to take our Flight to SIMIKOT. Fortunately, weather God favoured us and entire crowd of almost 100 pilgrims was lifted to SIMIKOT the same day. There were around 7 - 8 Airlines operating flights from NEPALGANJ to SIMIKOT. Aircrafts deployed on this route were small with capacity to carry 18 passengers with two pilots and no cabin crew. One pilot appeared to be a senior and experienced but the other looked to be a rather young hand. During our flight, we came across two similar type of Aircrafts returning from SIMIKOT. These Aircrafts were flying approx 500 meters apart from us and looked so tiny. Sight generated bit of amusement and awe. Initially, we had doubts about these small flying machines but were set right once landed safely at SIMIKOT and in a stipulated time of 45 mimutes.

SIMIKOT is in the Western extreme of Nepal and HQ of HUMLA District in Karnali Zone. It is closer to Dharchula in the Kumaon region of Uttarakhand. Eight days walk from the Terai region of Nepal and not connected by any road communications. One has to either walk for eight days from foothills or take a Flight from NEPALGANJ to reach SIMIKOT. It is a beautiful place located at an altitude of 9,200 feet and nestled in complete wilderness. If one truly wants to be away from the daily hum drum of city life, maddening crowds all around and away from all types of pollution, SIMIKOT is the place to rest for a short or long stay and rejuvenate. Also, a short visit to SIMIKOT will definitely take you back at least by one hundred years, put you at complete ease in the lap of Mother Nature and make you learn to survive with just bare essentials. I heard from local residents of this place that inhabitants of number of nearby villages still wear overcoats made of animal hides during Winters to protect themselves. Place gets covered with 3 to 4 feet of snow during Winters and becomes inaccessible even by Air from November to mid May each year. It has a rather small and dangerous airstrip to land. It is said this air strip features in one of the most dangerous airstrips to encounter. We experienced it ourselves; it must be true. Same air strip is used for concurrent landing and take off by the Helicopters too. It has been common for the moving rotor of the Helicopter touching a fixed wing aircraft parked nearby, thus damaging one or both of them. Fortunately, we were scheduled to fly out to HILSA by helicopters the same day. Also, there was not enough accommodation available at SIMIKOT to house 150 to 200 pilgrims of onward and return journey for the night.

I had read the place name SIMIKOT in a book named 'WANDERINGS IN THE HIMALAYAS' written by SWAMI TAPOVAN MAHARAJ. He was an enlightened Saint of 20$^{th}$ Century who spent his life time in UTTARKASHI and area North of GANGOTRI. Since he used to stay at a place called TAPOVAN, ahead of GANGOTRI, for major part of the year, he was better known as SWAMI TAPOVAN MAHARAJ. He undertook numerous treks to KAILASH and in one of them traversed through SIMIKOT, sometime in 1930s. He vividly explained the travails and tribulations of going through this undeveloped and rather primitive area almost a century ago.

Numerous Helicopter Companies had deployed their flying machines for lifting the pilgrims from SIMIKOT to HILSA and back. Every one of us wanted to be at HILSA earliest. Each helicopter could lift only five pilgrims with their accompanied baggage at a time. We were lifted in two Helicopters lifts and dropped at HILSA. Helicopter journey from SIMIKOT to HILSA was very beautiful. Mountains underneath were very high, steep and number of them appeared to be conical. There was hardly any inhabitation beyond SIMIKOT and same kept decreasing further as we neared snow covered Great Himalayas. We neared Tibetan Plateau in about 15 minutes and our Helicopter lowered down in a small valley. We could notice a helipad and numerous Helicopters were busy flying in/out. This was rather small a place and bit of anti-climax to our expectations. Altitude of HILSA was around 14,000 feet and we had to spend a night here. There was barrack type of accommodation which had no beds and cotton mattresses had been spread on the floor and quilts provided to cover ourselves. It was quite cold and high altitude effect further added to the discomfiture of pilgrims. Imagine, we all were at Mean Sea Level at NEPALGANJ in the morning and had been lifted to this altitude by the evening without having gone through any acclimatisation. Not so favourable a situation for many of us. Fortunately, we had a Maharashtrian pilgrim among us who was a practicing Doctor in Ireland. He took control and administered high altitude medicines to all needy persons. In fact, Tour Operators also had made appropriate arrangements for this eventuality. Arrangements for our night stay and food at HILSA were fairly satisfactory. As also, we did not expect city-like arrangements at such difficult places. It was evening by the time we settled down. Beauty of Sunset, with golden hue of sun light falling on the barren hills around, was beautiful. River KARNALI was flowing just close by. It enters Nepal from Tibet at HILSA. There was a temporary bridge linking HILSA with Chinese Immigration Office other side of the River.

Next day we all, a combined group of almost 100 pilgrims, crossed the bridge over River KARNALI and entered Tibet. Chinese immigration office was not far. They did their work without creating any hassles. Chinese buses had been lined up and took us to a famous place called TAKLAKOT, approx 3 km away from Chinese Immigration Office. It was mid day by the time we reached a Hotel where arrangements for our stay had been made. Place was like a 3 STAR Hotel. TAKLAKOT is located in the PURANG County of Tibet and at an altitude of approx 14,200 feet. Fairly good arrangements had been made by the Tour Operators in conjunction with local Chinese authorities. Our accommodation arrangements had been made by the Chinese authorities where as food part taken on by our Tour Operator. It was typical North Indian food, in plenty and served as buffet to all. TAKLAKOT has been developed by Chinese authorities as a model township in the wilderness, for all foreign pilgrims to see and acknowledge the development made by China in Tibet. We all once again crossed KARNALI River to visit the TAKLAKOT Township. View of KARNALI River flowing down with blue water giving the clear description of blue sky over our head was very beautiful. TAKLAKOT looked like any freshly developed township, neat and clean, broad roads, triple storied shopping complexes on either of the roads to take care of general needs of the pilgrims and other locals which were very few. This was the first place where from we could make use of high speed internet and other communications. Local population on the street was almost negligible. We spent approx two hours in the market, purchased few items and returned to our Hotel other side of KARNALI River. One young lady, in her mid 30s and from Mumbai, developed severe high altitude sickness. She had lost control over her body organs and had to be evacuated down to SIMIKOT. It was sad; she came all the way up to TAKLAKOT and had to return without having fulfilled her wish to pay obeisance at KAILASH – MANSAROVAR. Good that she returned safely, anything could have happened if not timely evacuated to lower heights. We were accompanied by a group of 30 young women from Aurangabad, Maharashtra. They had no male members accompanying them. Even their Tour Coordinator was a female. It was nice to see our women undertaking such arduous journeys to so far off places and independently.

A Chinese Government appointed Guide took control of the entire lot of pilgrims from TAKLAKOT onwards. He was to manage all our pilgrimage related movements hereafter for the entire duration of our stay in Tibet including going around MANSAROVAR Lake and PARIKRAMA of KAILASH. Next day morning we all got into three buses and started our move for MANSAROVAR Lake, our next night halt. It was at a distance of approx 20 km from TAKLAKOT. En route we halted at RAKSHASH TAL, a huge salt water lake on West of road. It is an endorheic lake, meaning there is no out flow of water from that basin. Hence, its water is not potable. We were instructed to keep safe distance from this lake as the lake was considered inauspicious. Sight of this lake was beautiful. Also, it was from here that we were able to have our first view of PIOUS KAILASH, an oval shaped pyramidal structure fully covered with white snow/ perennial ice. Just adjacent to RAKSHASH TAL and on to its East was MANSAROVAR Lake and the road passed between the two lakes. South of Lake MANSAROVAR was famous mountain named GURLA MANDHATA which remains covered with snow throughout the year and continuously feeds water into MANSAROVAR Lake. It is pear shaped and highest fresh water lake of the world. Its widest part is 26 km and has a total circumference of approx 90 km. It covers a total area of 412 sq km and located at an altitude of approx 14,500 feet. Lake water is crystal clear and glimmering blue under the sunshine; one can see through the water for 5 to 7 meters inside. Its maximum depth has been estimated to be 70 meters. We reached our place of night stay on the bank of this Holy Lake. Since we had come further close to KAILASH PARVAT, its view had become much clearer and looked very beautiful. Numerous barrack type small huts had been made for the night halt. Wooden cots with cotton mattresses and quilts had been provided which were quite sufficient. There was a small Gompa, named JI WU, located on a small hillock, just 200 meters away from our place of stay. In the afternoon, we were again embussed and taken on 90 km long trip going all around MANSAROVAR Lake. Colour of its water was changing from blue to green to blue depending upon the direction of sunlight falling on it. We were halted at a designated place generally used for carrying out Havan / Pooja and safe for bathing by the pilgrims. We all 10 members of the group carried out Havan together and took a Holy dip in the water of Pious Mansarovar, a life time dream of any

Hindu. It was a great occasion, an unforgettable and most pious event of our life. A dream come true. We all considered ourselves truly blessed.

The view of KAILASH PARVAT, 20 km away from the lake, was superb. Bluish green water of the Lake, as it appeared at that time of the day, lying between us and KAILASH , added immense beauty to the spectacle spread in front of us. It is simply inexplicable and can only be experienced by an individual. Memory of that spectacle is permanently etched in my memory. Here we saw a local person, in his mid 30s, calmly busy covering the entire 90 km circumference of the Pious Lake by doing SHASHTANG PRANAM. In fact, he was to continue likewise to KAILASH and further do PARIKRAMA around in the same manner. Immense amount of faith, devotion and dedication is required to undertake such arduous efforts. We returned to our place of night halt at the Western edge of the Lake.

Before proceeding on this pious journey, my friends back home had asked not to miss the sight of a light which emerges on Eastern Sky above the Lake at around 2 am on every POORNIMA ( full moon night ) for a few minutes and then merges in the Lake. Fortunately, It was the full moon night. At 11 pm in the night, our entire group came out in the open to watch the spectacle of emergence of unknown light on Eastern Sky. We waited for an hour or so but felt too tired and sleepy to keep awake for 2 – 3 more hours and returned to our barracks. Those who kept awake neither confirmed nor denied seeing the unknown light during the previous night. I suppose once faith in its occurrence plays an important role.

Next was our move further to a place called DARCHEN, 20 km due North of MANSAROVAR. KAILASH PARVAT was nearing with every km of distance going behind and its sight becoming clearer and clearer. We all were getting truly mesmerised by the very sight of KAILASH; and, so near to us. DARCHEN was still few km distance away. Our full attention was on the sight of KAILASH. Suddenly, we noticed a figurine in the midst of perennial snow and on the right bottom of the KAILASH Dome. In the very first sighting, it was not clear as to what it was and started getting clearer as we came further close to KAILASH. I moved ahead from my seat and took permission of our Guide to sit near the driver so as to have a better look ahead. I was amazed to see that the figurine looked like a female. This female figurine was seen on left bottom of the snow covered portion of the KAILASH Dome and was visible to all of us till we were 3 -4 Km away from DARCHEN. We took Photos and made video also of this occurrence. We all accepted this to be MAA PARVATI and bowed our heads in reverence .

DARCHEN is located on East – West Highway running across Southern Tibet and connects with AKSAI CHIN and other parts of Eastern Ladakh. It is an important township developed by China and has all essentially required facilities. A proper market place, even better than one at TAKLAKOT, had been developed. There were numerous Home Stay and small Hotels available and we all made fairly comfortably. Good food was provided all through. In fact Tour Operator had moved his entire Catering Staff from TAKLAKOT to DARCHEN.

DARCHEN was at even higher altitude, approx 15,500 feet. One does experience breathing problem, especially if not undergone proper acclimatisation which most of us had not. Here segregation of pilgrims wishing to undertake three days long PARIKRAMA all around KAILASH PARVAT is carried out and who do not wish to for some reasons are required to stay put at DARCHEN itself. Since myself, wife and Mr V S Kotwal and Mrs Jaishree Kotwal were in the higher age group, we decided to stay back and remainder members of the Group were to proceed on PARIKRAMA next day. Same day we all were taken to a place called YAM DWAR, approx 4 km from DARCHEN on way to KAILASH. All pilgrims irrespective whether going on PARIKRAMA or not, are permitted to go up to YAM DWAR and have a very clear and closest look at KAILASH. There is a small Gompa like structure made at this place and all pilgrims go around the Gompa as a mark of reverence. View of KAILASH PARVAT from YAMD WAR is truly fabulous. Entire Dome is covered with snow / perennial ice. Strange as it may appear, it looks as if 10–12 large steps have been cut on the rocky and perennial ice covered face of KAILASH PARVAT starting from almost bottom and going up to approx ¾ height of the Dome. At the end and left of the last step, is almost look-alike of a human face with masculinity in looks. With bit of imagination thrown in, this face has been accepted to represent SHIV PARMATMA. We all bowed down to offer our sincere and most respectful obeisance to the Deity. A once in life time opportunity to be face to face with SHIV PARMATMA in his abode and paying obeisance to Him directly is the greatest blessing one can have. We were truly blessed; grateful to Almighty to enable us to reach this place and stand right in front of Him and at His pious abode.

Next day, 85 out of total 100 pilgrims set out for PARIKRAMA around KAILASH. We four belonging to higher age group had decided to stay back; remainder six members formed part of the 85 pilgrims proceeded ahead for PARIKRAMA. We noticed that we four who decided to stay back were not alone. We had nice company of four hefty looking old Marwari ladies from Kolkata, defying the thinking that obese people should avoid taking the risk. One should have faith in the Almighty and leave the remainder to Him. What else does one need after having crossed 70 years of age!!

It is a tough three days trek to go around KAILASH PARVAT. On second day, one has to climb up to the height of 19,000 feet to get across for third day's march. Many people had to return after completing first day's march. Most of the pilgrims could not go beyond second day's march. Ultimately, only five out of 100 pilgrims who started, were able to complete the PARIKRAMA, rest all returned midway. Only one from our group of six pilgrims, named RAM BABU, could successfully complete the PARIKRAMA. Our youngest member, Nimit Tiwari, developed Pulmonary Odema at the height of 19,000 feet and was quickly brought down by the Guide accompanying him. We all thanked God Almighty for His kind blessings on Dear Nimit Tiwari. On fourth day of our arrival from MANSAROVAR Lake, we all pilgrims had again safely gathered at

DARCHEN and were ready for next move.

Next day we all commenced our return journey. En route, we had an hour long halt at PIOUS MANSAROVAR Lake. It had been snowing there. We all left for TAKLAKOT and then to HILSA for obtaining an early Helicopter lift to SIMIKOT. Lot of returning pilgrims were already stranded there and we had to wait for almost three hours at the dusty helipad for our turn for Helicopter lift. We all safely reached SIMIKOT by 3 pm. Finding a suitable accommodation for all of us became a problem. Somehow, our Tour Operator managed to put us in a Home Stay type of accommodation. There were four rooms and we all shared the accommodation. We four, self, wife, Mukesh Somani and his wife Anju Somani shared one room which had only three beds. There was three hours of day light still available. Most of us visited a nearby old temple of SHIVA and paid our obeisance. There was a small market, most of the shops hardly had any stores and were in dilapidated state. Local Market remains open during the Pilgrim Season only which lasts from June to October / November each year. Districts of HUMLA, JUMLA and DOTI which form part of KARNALI Division of Western Nepal have remained undeveloped. One has to bear this in mind before we raise our level of expectations for administrative comforts at such far flung places. Development begins with opening of road communications and other developmental activities follow soon after. The Home Stay accommodation we stayed belonged to a THAKALI ( Business Community person ). He was in his early 30s and had studied in Kathmandu till graduation. He was intelligent and quite well versed with political, economic and administrative affairs of the country. In fact, he knew few Senior Officers of Indian Army by name who had been Military Attaches in Kathmandu. His house-hold and a community kitchen were located at the ground floor and we all visitors stayed in the accommodation at First Floor. It was a fairly satisfactory arrangement for the short stay.

Sometime in mid of the night, Anju Somani, wife of Mukesh Somani with whom me and my wife Kamlesh were also sharing the accommodation for the night, developed severe stomach ache.

A stage came when it became unbearable. There was no medical aid available anywhere in the near vicinity. Her husband became helpless as he could do just nothing to help her. So were me and my wife and we all three kept telling Anju Somani the whole night she will be alright soon. After 2 -3 hours of suffering, she started improving and was alright by the morning. Both of them were able to catch an early flight to NEPALGANJ so as to consult some Doctor in an earlier time frame. Mr V S Kotwal and his wife also were able to accompany them to NEPALGANJ in the same flight

There was lot of rush at the Airport and it was around 1 pm that we were able to catch the flight. Available Flight was routed to drop us at SURKHET as NEPALGANJ Airport had become congested of numerous flights landing / taking off that day. SURKHET was approximately 110 km North of NEPALGANJ. Aircraft we were allotted was a rather tiny, just sufficient to accommodate 6 passengers excluding 2 Pilots. It was really looking like a Toy Aircraft. We again had serious doubts before it valiantly took off from the Airstrip without any hassles and in an hour's time dropped us at SURKHET Airport. Great experience, indeed. We soon hired a worn out Taxi which somehow brought us to NEPALGANJ. After carrying out final coordination with our Tour Operator and remainder members of our Group, we four, to include myself, my wife Kamlesh, Dr Anupam Goyal and Nimit Tiwari hired a Taxi from NEPALGANJ to go to Lucknow and around 4 pm started our journey back home. En route, we changed our plan and instead of going to Lucknow and then taking a train to Delhi next day, we coordinated with the Taxi Driver to drive us straight to Delhi. It was a long drive for almost 9 hours in the night. We reached Delhi at half past mid night and safely reached our home in Noida at 1.30 am in the morning.

This was our truly great visit, a very memorable one and a life time achievement. The Joy of having visited KAILASH PARVAT, seen it in person, paid our obeisance to SHIVA in its abode and bathing in the water of most PIOUS MANSAROVAR Lake, cannot be put writing nor can it be explained to any one by any means; no Ted Talk or Power Point Presentation can ever explain you, in true sense. IT HAS TO BE EXPERIENCED. We had been blessed by the Almighty to go on this visit to PIOUS KAILASH PARVAT and MANSAROVAR Lake and are ever grateful to SHIV PARMATMA. Unfortunately, we lost one member of our group, BASANT JHA, who died of CORONA pandemic in March 2021 at his home in Ajmer. May God grant SADGATI to the departed soul.

*KAILASH PARVAT from YAMDWAR.*

*KAILASH PARVAT from YAMDWAR.*

*Most Beautiful View of KAILASH PARVAT and MANSAROVAR.*

*Pilgrims on the Banks of MANSAROVAR.*

*Self and Wife with KAILASH on the Skyline.*

*Self at DARCHEN.*

*KAMLESH at YAMDWAR with KAILASH in the Background.*

*With a Tibetan Guide at MANSAROVAR.*

*RAKSHASH TAL near MANSAROVAR.*

*We the Pilgrims at SIMIKOT in Nepal.*

*Pilgrims at YAMDWAR.*

*Helicopter Melee at HILSA near TAKLAKOT in TIBET.*

## WITNESS TO A DIVINE SPECTACLE BEYOND SHREE BADRINATH DHAM

It is difficult to define the word 'My Spiritual Experience'. This experience occurred to me in pious surroundings of 'Great Himalayas', beyond Shree Badrinath Dham. I have termed this divine experience as 'My Spiritual Experience' as I felt divinely elevated far beyond the Galaxies visible to a human being by any means.

I belong to a humble, middle class, agriculturist family and provided a fairly good education by my parents. I served in Army for 35 long years and, thereafter, 8 - 10 more years in various other organisations in civil life. By then I had completed all my socials obligations as a house holder and finally settled down in Noida. It was at this time I started getting drawn towards the realm of Spirituality in true sense. WHO AM I ? WHERE FROM HAVE I COME ? WHAT IS THE PURPOSE OF THIS LIFE? Many more such like questions were now thronging my mind. I started visiting various Ashrams in Rishikesh, Haridwar, Dehra Dun and Dharamsala (HP ), collected numerous basic books on the subject for carrying out my Self Study, attended numerous Retreats on basic Scriptures conducted by learned SWAMIJIS in various Ashrams and thus made a fairly serious attempt to fully immerse myself in to the subject. Self and wife were always together in this sincere and pious effort and carried out Self Study together. In that, we both would sit together in a room, I shall read certain SHLOKAS and their explanation given by the learned Swamiji in the book and then discuss. There was never any hurry in doing so. First reading of both the Volumes of VIVEK CHOODAMANI and SHREEMADBHAGVAD GEETA took us one full year. Subsequent revisions took lesser time. It was a satisfying arrangement to somewhat comprehend the basics of Spirituality. In any case, a beginner always finds difficult to be drawn toward Spirituality in his journey toward the Unknown. It is not an attractive subject for those who believe in 'SEEING IS BELIEVING'.

In June 2016, self, and wife went on a week-long pilgrimage to holy shrine of Shree Badrinath Dham. Month of June is a full pilgrimage season for the CHAR DHAM YATRA and pilgrims from far and wide visit YAMUNOTRI, GANGOTRI, KEDARNATH, BADRINATH and HEMKUND SAHIB shrines. It was this pilgrimage that I was reminded of a Trek from SHREEBADRINATH DHAM to SATOPANTH TAL, nestled deep in Great Himalayas that existed since ancient days. This Trek is known as SWARGAROHINI, literally meaning 'CLIMB TO HEAVEN'. All five GREAT PANDAVA WARRIORS, accompanied by QUEEN DRAUPADI, had undertaken this pious trek to wash away the sins of having killed their own kin, KAURVAS, in the battle of MAHABHARATA , as also to attain MOKSHA ( Salvation ). In childhood, my parents used to narrate story of this 'Trek to Heaven' and it remained etched in my sub conscious mind as a childhood fantasy and a sincere desire to undertake some day. Student life, followed by service in Army and other family constraints lowered the priority, yet it always remained in my subconscious mind. It started taking shape into a reality once I visited Shree Badrinath Dham in 2016. I noticed numerous shops selling Videos of various pilgrims / trekkers going on this trek. This further strengthened my resolve to undertake this Trek. I was already on wrong side of age, 69 years to be precise, to proceed on such an arduous Trek. If I have to do this, it has to be done soon.

SWRGAROHINI Trek starts from Badrinath Dham and goes up to Satopanth Tal, a triangular shaped emerald colour lake nestled at the base of CHOUKHAMBA Mountain forming part of Gangotri group of mountains. It is a 54 km long to and fro journey and completed in eight days. It is mandatory to undergo essential high altitude acclimatisation for a minimum duration of four to six days before undertaking this arduous journey in deep Himalayas. We both engaged ourselves in study of basic scriptures and regular meditation to drive away irrelevant and impure thoughts and thus make sincere effort to cleanse our mind and intellect.

Also, we were in the process of undertaking a holy pilgrimage to KAILASH and MANSAROVAR in Tibet, year later in 2017 and preparations were afoot for that alongside. All these were driving us to immerse ourselves fully well in all possible pious activities. During early September 2017, we went on a week-long visit to KASHI and paid our obeisance at the feet of VISHWANATH, the Presiding Deity and visited numerous GHATS on the banks of Holy Ganga in existence since times immemorial. All these activities were sub consciously adding to the process of cleansing of our 'Inner- Consciousness' of all non-essentials and preparing for the long Spiritual journey ahead.

After four months of our pilgrimage to Badrinath, I was informed by the Tour Operator at Joshimath to reach Badrinath by 18 September 2016 to undergo necessary high altitude acclimatisation and equip myself for SWARGAROHINI Trek to Satopanth Tal. My heart was filled with immense joy as I will now be able to fulfill my childhood dream. I took the first possible Bus from Noida and for Haridwar and arrived at Badrinath in two days. I stayed with GARHWAL SCOUTS Camp and commenced higher stage acclimatisation in right earnest from 19 September 2016 onwards. I used to follow right bank of ALAKANANDA River, walk 4 - 5 km towards VASUDHARA WATERFALLS and return to Badrinath, thus covering a distance of 8 – 10 km every day.

All this which has been narrated above is a mere prelude to My Spiritual Experience. It has a direct bearing to assist the reader in comprehending what was happening in the body, mind and intellect of the EXPERIENCER, what factors directly or indirectly may have contributed in the occurrence and what was the emotional state of the person involved. Please remember that such an occurrence cannot be explained in spoken or written forms or given as Power Point Presentation or TED TALK. These can only be EXPERIENCED by a person. Still, a sincere and earnest endeavour has been made in bringing the Reader somewhat close to understand it.

It was 21 September 2016 and third day of my acclimatisation schedule. It was almost end of September, rain clouds had receded, weather was absolutely clear and not even a speck of cloud could be seen in the sky. No smoke or dust pollution, hence, sky all around was blue and breathtakingly beautiful. It was during the course of that day's walk towards VASUDHARA WATERFALLS that I went through a DIVINE EXPERIENCE which was SPIRITUAL in nature. I had walked along the track for about 4 or may be 5 five km and VASUDHARA WATERFALLS was still one km or so ahead. I noticed there was no one ahead or behind me on the track for whatsoever distance I could observe. I was all alone in the wilderness. There were massive, high, barren and rather steep rising mountains, almost 70 to 80 degrees, on either side of mine. On to my left and just 100 meters from the track, I could well hear the sound of Alakananda River flowing down toward Badrinath. Also, I noticed a pair of black coloured Hill Myna occasionally flying over my head in the sky and making sound in their own language as if chatting with each other and fully enjoying the silence of Great Himalayas. Other side of Alaknanda River was the track leading to Satopanth Tal which I was to traverse soon. I could see 3 – 4 glaciers rolled down and had effectively obliterated all signs of track for a long distance. I could still notice few Trekkers and Coolies wading through the glaciers and moving ahead towards Satopanth Tal.

It was around 11 am. VASUDHARA WATERFALL was still approx one km ahead. It was all stunningly quiet around me. For some unknown reasons, I felt like halting for a few minutes wherever I was standing at that very moment. I left the track and moved 3 – 4 steps toward Alakananda River. A sudden thought came to my mind to offer prayers to the God Almighty thanking Him for giving me an opportunity to be in such pious and beautiful surroundings.

I faced toward MAA ALAKANANDA, folded my hands in reverence, closed my eyes and prayed to SHREE BADRIVISHAL, the reigning deity of Shree Badrinath Dham, NARAYAN Himself. My prayers lasted for 2 – 3 minutes. Once I had finished my prayers, I had a very strange feeling and did not know what was happening to me. My eyes were still closed and it was all dark in front of my eyes. It was at this moment that I felt elevated to a different world all together. In that state, I had no physical body of mine nor could I experience my body anymore. However, I could clearly see all around me, far and wide but not my own self. It appeared as if I have given DRISHTHI – 'Power To See All Around' – without having my physical body. It was sort of a strange experience which I had never undergone anytime in the past.

I was not on this Earth but far beyond it; in fact far above the Sky usually seen by me. Entire Sky now visible was much below me and I was much above the Sky. There was no Sun or Moon or any other Planets in the Sky. It was almost a dark Sky, one generally observed on the night of AMAVASYA and yet so beautifully lit up by astral light. I could see billions and billions of Stars twinkling in the Sky below me and all around. I very vividly remember there were no Stars above me. Entire Sky was below me. Astral light had made it possible for me to clearly observe billions of Stars twinkling below me. Scene below me was similar to the night of AMAVASYA in Hindu Calendar when there is almost no Moon in the Sky. Most of the Stars were small in size with a sprinkling of few big Stars also. I could see no Moon, SAPTARISHI or any other Star which I could recognise. Also, there was no Milky Way in the Sky which otherwise is so clearly visible during any dark night. I was definitely not day dreaming. I was very much fully awake and standing on my two feet. I was simply thrilled to see this unbelievable spectacle unfolded before me. What a blessing; standing on the bank of Pious Alakananda River - almost at its UDGAMSTHAL ( Point of Origin ), in the midst of High Himalayas, no human being around and witnessing the grand spectacle unfolded before me.

So far I had not recognised any Star in the Sky below me. Now, I noticed a bigger Star in front, little on to my right hand side and appeared bit familiar to me. I suddenly felt I know this Star as planet Earth. It appeared much closer and was shining much brighter as compared to other billions of Stars twinkling in the sky. Its size appeared equivalent to that of a Tennis Ball and, as said earlier, was brightest of stars visible to me at that very moment..

I observed there was now something protruding upwards from top of this Star and reached almost 3/4 of the distance between me and Planet Earth. This protrusion was not there on this Star when I initially recognised it as Planet Earth. For most part of its initial protrusion towards me, I could only see a hazy outline. I looked at the top part of the protruded structure more intently and recognised it as a Temple. Architecturely, the top portion of this structure looked somewhat similar to the Temples of Kedarnath, Tungnath or any other temples generally found in the hills of Uttarakhand. There was a SHIKHAR atop this Temple like structure but did not appear to be made of any metal. I could not see any doors to this Temple as those would have been much lower and not visible to me. During this part, Earth continued to be visible to me. After some time the vision of this temple like structure disappeared from my view while Earth continued to shine as bright as before.

Now, I noticed a HALO around Earth which was not there till now. This HALO appeared to be shining even brighter than Earth. Width of this HALO was about 1/4 or may be 1/5of the diameter of Earth as visible to me. Suddenly, this HALO started rotating around Earth. Movement was at uniformly slow pace and in clockwise direction. This rotary movement of the HALO continued for approximately little over one minute and then stopped. I regularly took an all round view of Starry Spectacle below me. Billions of stars were shining as before and presenting a truly magnificent and divinely sight. I again focused my view on to the Earth which was appeared so small in the vast Sky before me. After a gap of 10-15 seconds, I noticed HALO had again started moving in rotary direction around Earth. This time the movement of HALO was in anti-clockwise direction and with same uniformly slow speed as was in clockwise direction earlier. Anti-clockwise movement of the HALO lasted for approximately one minute, little lesser time as compared to clockwise direction. Thereafter, all of sudden, the rotary movement of HALO the Planet Earth stopped. With that, the Starry Spectacle also disappeared from my vision. Now, it was all dark in front of me, as dark as when I had closed my eyes to offer my prayers to SHREE BADRIVISHAL in the very beginning. Above mentioned EXPERIENCE may have lasted for 9 or 10 minutes.

I slowly opened my eyes. I still had my hands folded and standing at the same very place facing Alakananda River. Sky was still bright and Sunny as before. I could see a pair of Hill Myna flying and frolicking above my head. I could even hear their sweet voice in my otherwise hearing impaired ears. There was no human being in sight anywhere around. I was so overjoyed, my happiness knew no bounds. I was so much elated after witnessing the Starry Spectacle. It is beyond any description. It was the grace of SHREE BADRIVISHAL and MAA ALAKANANDA that enabled me to witness this spectacle, a sort of VISION for me. I continued to stand at the same very spot, in the same posture with folded hands and kept pondering over what God Almighty has shown to me. I decided to return to my Base at GARHWAL RIFLES Camp at Badrinath from that very place itself.

I discussed this occurrence with some knowledgeable persons at Joshimath and area around. One person stated that such occurrence sometime occur while undertaking a SWARGAROHINI Trek to SATOPANTH TAL, holy pilgrimage to Kedarnath or other prominent Temples. I was eager to know much more on what I had witnessed. This could be possible if I meet some people who are learned and advanced in this field.

In March 2017, I got an opportunity to visit CHINMAYANANDA ASHRAM at Dharamsala (HP) where I attended a 20 days long Retreat on VIVEK CHOODAMANI (Crest Jewel of Intellect). Retreat was conducted by a learned and spiritually accomplished SWAMI JI. During the Retreat, I sought a personal meeting with the Respected SWAMIJI and discussed occurrence of this EXPERIENCE. SWAMIJI stated there are many places in the deep interior of Himalayas where numerous distinguished and enlightened RISHIS have undertaken TAPASYA and SADHANA for a protracted period in the past. It may have happened hundreds or even thousands of years ago. Because of their TAPASYA and SADHNA, that specific place has become permanently energised. If a person with pure 'Inner Consciousness' happens to be there, he may undergo such an 'EXPERIENCE' or 'VISION'. I further enquired from Respected SWAMIJI if there was any message meant for me in this experience and how should I proceed hereafter. He stated that these experiences are one time occurrences and carry no message as such. He further clarified that I had pure 'Innerself' (ANTAHKARNA)at the time of this EXPERIENCE and may continue in the spiritual journey ahead, if I so desire. Also, we are regular visitors to SIVANANDA ASHRAM at Rishikesh. In June 2018, I discussed this experience with another very learned and accomplished SWAMI JI. He stated that these occurrences are one time happenings in a pure 'Inner Consciousness', do not signify anything much and advised me to continue progress ahead in the Spiritual Journey.

I am truly grateful to SHREE BADRIVISHAL for enabling me to become a witness to such a grand Starry Spectacle of the Universe. It was all His Grace. We are forever grateful to you, SHREE BADRIVISHAL.

# ABOUT THE AUTHOR

Brigadier R V Singh obtained his education from D A V College, Dehra Dun, a city which had a large number of Military training establishments, ie, IMA, RIMC and numerous Gorkha Training Centres to influence and motivate younger generation to join Armed Forces. Like most of his College compatriots of those days, he also joined Army and was commissioned in First Battalion of the Fifth Gorkha Rifles ( (Frontier Force) in September 1970. He served with his Battalion for 7 continuous years before going on a Grade 3 appointment to a Brigade HQ. In those first 7 formative years he participated in Indo-Pak War of 1971, served in J&K, Arunachal Pradesh and Mizoram, as usual of any other Infantry Officer. Later he had two tenures in High Altitude Areas in North East. He did numerous Staff and Instructional appointments and attended Staff Course at DSSC, Wellington (T N) and Higher Command Course at College of Naval Warfare, Mumbai.

He had a good mix of Regimental and Staff assignments throughout his 35 years long career in Army. He either participated in various incidents himself or witnessed them from close vicinity. He has written his experiences in anecdotal form as the incidents unfolded before him at varied place and time. A first person account of own experiences and written in a rather straightforward, unbiased and impartial manner. Should make a good leisurely reading for all.

www.ingramcontent.com/pod-product-compliance
Lightning Source LLC
Chambersburg PA
CBHW072249210326
41458CB00073B/914

*9789393499363*